Green Energy and Technology

Climate change, environmental impact and the limited natural resources urge scientific research and novel technical solutions. The monograph series Green Energy and Technology serves as a publishing platform for scientific and technological approaches to "green"—i.e. environmentally friendly and sustainable—technologies. While a focus lies on energy and power supply, it also covers "green" solutions in industrial engineering and engineering design. Green Energy and Technology addresses researchers, advanced students, technical consultants as well as decision makers in industries and politics. Hence, the level of presentation spans from instructional to highly technical.

Indexed in Scopus.

Indexed in Ei Compendex.

Paulo Sérgio Brito · João Rafael Galvão ·
Bruna Rijo · Cecília Pedrero · Filipe Neves ·
Henrique Almeida · Joel Oliveira Correia Vasco
Editors

Insights into Energy Trends

Selected Papers from ICOWEFS 2024

Springer

Editors
Paulo Sérgio Brito
Portalegre Polytechnic University
Portalegre, Portugal

João Rafael Galvão
School of Technology and Management
Polytechnic University of Leiria
Leiria, Portugal

Bruna Rijo
Portalegre Polytechnic University
Portalegre, Portugal

Cecília Pedrero
Portalegre Polytechnic University
Portalegre, Portugal

Filipe Neves
School of Technology and Management
Polytechnic University of Leiria
Leiria, Portugal

Henrique Almeida
School of Technology and Management
Polytechnic University of Leiria
Leiria, Portugal

Joel Oliveira Correia Vasco
School of Technology and Management
Polytechnic University of Leiria
Leiria, Portugal

ISSN 1865-3529 ISSN 1865-3537 (electronic)
Green Energy and Technology
ISBN 978-3-031-83810-1 ISBN 978-3-031-83811-8 (eBook)
https://doi.org/10.1007/978-3-031-83811-8

This Springer imprint is published by the registered company Springer Nature Switzerland AG
The registered company address is: Gewerbestrasse 11, 6330 Cham, Switzerland

Scientific Committee

Abel Rodrigues, INIAV/Research Institute, Portugal
Ahmad Pourmovahed, Kettering University, Michigan, USA
Alfeu Sá Marques, Coimbra University, Portugal
Ana Gomes, Porto Catholic University, Portugal
Ana Vale, School of Science IT Sligo, Ireland
Annukka Pakarinen, HAMK University, Finland
Antonio Macías, Extremadura University, Spain
Bruna Rijo, Portalegre Polytechnic University, Portugal
Carla Ferreira, ESAC/Coimbra Polytechnic, Portugal
Carlos Henggeler, INESCC/Coimbra University, Portugal
Catarina Nobre, Portalegre Polytechnic University, Portugal
Cecília Pedrero, Portalegre Polytechnic University, Portugal
Daniel Gutierrez, Univerdidad Técnica Manabi/Equador
Eduardo Álvarez, Oviedo University, Spain
Eliseu Monteiro, FEUP/Porto University, Portugal
Fausto Freire, CIE/Coimbra University, Portugal
Fernando Lindon, FCT Lisbon University, Portugal
Fulvia Chiampo, Politecnico di Torino, Italy
Hassan El Bari, Kenitra University, Marocco
Helena Bártolo, Polytechnic University of Leiria, Portugal
Henrique Almeida, Polytechnic University of Leiria, Portugal
Howard Dryden, Scotlnad University, UK
Jerónimo Cortés, CICTE/Extremadura Research Center, Spain
João Pina, FCT/Nova Lisbon University, Portugal
John Cosgrove, Limerick Institute of Technology, Ireland
Jorge M. G. P. Isidoro, Algarve University, Portugal
José Lobo, Extremadura University, Spain
Kirsi Knuuttila, JAMK University, Finland
Lisiane Librelotto, UFSCatarina, Brasil
Lucélia Hoehne, Universidade do Vale do Taquari—UNIVATES, Brazil
Luís Loures, Portalegre Polytechnic University, Portugal

Organising Committee

Ana Carolina Assis
Portalegre Polytechnic University

Bruna Rijo
Portalegre Polytechnic University

Catarina Nobre
Portalegre Polytechnic University

Cecília Pedrero
Portalegre Polytechnic University

Filipe Neves
Polytechnic University of Leiria

Flávio Craveiro
Polytechnic University of Leiria

Henrique Almeida
Polytechnic University of Leiria

Joel Vasco
Polytechnic University of Leiria

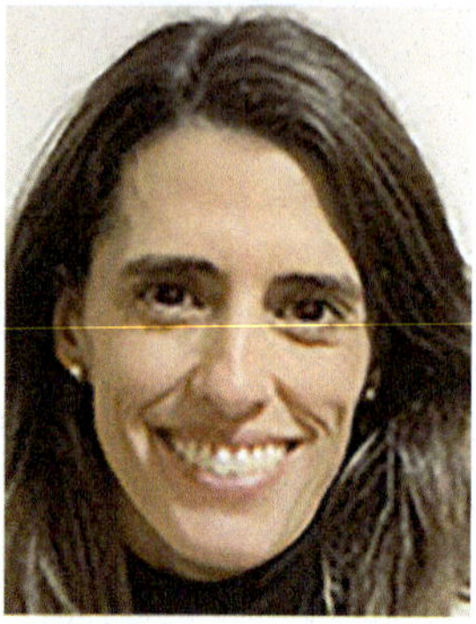

Liliana Ferreira
Polytechnic University of Leiria

Pedro Gala
Polytechnic University of Leiria

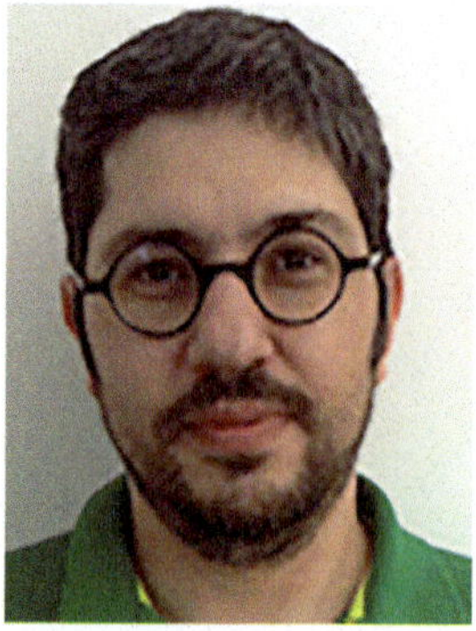

Pedro Monteiro
Portalegre Polytechnic University

Ricardo Gomes
Polytechnic University of Leiria

Roberta Panízio
Portalegre Polytechnic University

Sandra Mourato
Polytechnic University of Leiria

Santa Margarida Santos
Portalegre Polytechnic University

Vânia Ribeiro
Polytechnic University of Leiria

Conference Co-chairs

Paulo Sérgio Duque de Brito
Polytechnic of Portalegre, Portugal

João Rafael da Costa Sanches Galvão
Polytechnic of Leiria, Portugal

Conference Awards

The ReadyToPub Awards will be given to the best student oral communications and the best student posters (general and female participants). The main aim of the ReadyToPub Awards is to motivate and encourage young researchers to prepare their data and, with the help of ReadyToPub, attempt to publish papers in international journals with a high impact factor.
The evaluation of the communications will be based on five criteria. The awards consist of a credit worth €250 (best oral communications) and €150 (best posters) for use in services provided by ReadyToPub. The winners will be announced at the conference closing ceremony.

Best Student Oral Presentation

Author: Eduardo Costa Camilo
Title of the presentation: Future-proof a mediterranean soup

Best Female Student Oral Presentation

Author: Santa Margarida Santos
Title of the presentation: Characterization of refuse derived fuel pellets and feasibility assessment of their application in CO_2 gasification

Best Poster and Best Female Student Poster

Author: Sara Hassan
Title of the poster: The short-term effect of prescribed burning on aboveground biomass, and soil properties

Award Sponsor

readytopub
AUTHOR SERVICES PROVIDER

Conference Sponsors

The conference co-chairs and organising committee wish to acknowledge the support and sponsorship given in the organisation of the ICOWEFS 2024—International Conference on Water Energy Food and Sustainability, held at the Polytechnic Institute of Portalegre, Portugal.

Official Sponsors

POLITÉCNICO DE PORTALEGRE

POLITÉCNICO DE LEIRIA

ICoWEFS is Organized in Cooperation with

inesc 30 anos · coimbra

VALORIZA

MED
MEDITERRANEAN INSTITUTE FOR AGRICULTURE, ENVIRONMENT AND DEVELOPMENT

CHANGE
Global Change and Sustainability Institute

Event Sponsors and Exhibitors

Preface

The 4th edition of International Conference on Water Energy Food and Sustainability (ICoWEFS 2024), taken place in Portalegre (May 8–10, 2024) Portugal. The event sought to foster innovation and exchange knowledge in the water-energy-food nexus, embracing the Sustainable Development Goals (SDGs) of the United Nations, building a better future for all bringing together leading academics, researchers, and industrial experts.

The conference expects to foster networking and collaboration among participants to advance knowledge and identify major trends in the above-mentioned fields, in particular Energy. The production of energy and materials from biomass and waste materials is especially favored over fossil-based feedstocks since it has almost neutral net carbon dioxide (CO_2) emissions is one of the main objectives of the Conference.

The 2024 edition of ICoWEFS was a real contribution to the current effort of promoting ways of transferring knowledge by strengthening academic and industrial cooperation, which together can develop innovative and sustainable solutions for this energy transition.

This edition had the submission of more than 70 papers, with authors from several nationalities and with the collaboration of several keynote speakers, who enriched the debate and evolution of the scientific areas, specifically Energy, Sustainability, Water and Food, namely Agrifood. This book publication is only focused on the papers within the scope of Energy and is composed of 13 selected papers that the scientific committee considered relevant to the Energy domain.

Welcome to this 4th edition of the ICoWEFS 2023 Conference in Portalegre, Portugal.

ICoWEFS 2024

Portalegre, Portugal

Conference Chairs
Paulo Sérgio Brito, Ph.D.
João Rafael Galvão, Ph.D.

Keynote Speakers

Ana Luísa Fernando

Researcher at Mechanical Engineering and Resource Sustainability Center (MEtRiCS)
Ana Luisa Fernando is an Associate Professor at Universidade Nova de Lisboa, NOVA School of Science and Technology, Portugal. Researcher at MEtRiCS, Mechanical Engineering and Resource Sustainability Center, hosted by Universidade do Minho and Universidade NOVA de Lisboa. Graduated in Applied Chemistry (UNL), in 1990, with an M.Sc. in Food Technology/Quality (1996, UNL) and a Ph.D. in Environmental Sciences (2005, UNL).
She has been working with energy crops for more than 25 years, with a special interest in studies related with the Sustainability of energy crop production (use of marginal land; efficient use of water and mineral resources; environmental impact assessment studies to detect options for systems improvement). Parallel to those activities, she has also been working in the field of Food Technology and Safety, by testing natural compounds extracted from plants into biopolymers for food packaging or as additives for food preservation.
Supervision: 9 Ph.D. thesis (concluded), 10 Ph.D. underway, 121 M.Sc. thesis (concluded), 76 Graduation thesis (concluded).

She is also coordinating nationally the H2020 and HORIZON projects GOLD, MIDAS and IASIS. Author and co-author of several publications (h index = 35).

Increasing the sustainable production of biofuels—prospects of cultivating oilseed crops in soils contaminated with heavy metals

Abstract

The production of oil crops is an auspicious option for the partial substitution of fossil fuels in energy production. Several species can be cultivated for the use of biomass for energy production. Some new oil crops are promising due to their oil yields and quality. In addition to the potential for oil production, some species are capable of tolerating heavy metals from the soil. Thus, the present work aims to investigate the potential of Brassica carinata and Crambe abyssinica in soils contaminated with Zn (450 mg.kg-1, dm), Pb (450 mg.kg-1, dm), Cd (4 mg.kg-1, dm) and Ni (110 mg.kg-1, dm). B. carinata was the most productive crop (160 g/m^2), followed by crambe (130 g/m^2). Both oil crops can be considered tolerant to heavy metals in the study (tolerance index higher than 0.75). Yet, the oil content in the seeds was slightly affected by the level of contamination in the soil. Therefore, biodiesel production costs may increase because oil production per land area is affected by the level of contamination. Nevertheless, the production of oil crops in soils contaminated with heavy metals represents an opportunity to provide feedstock for the oleochemical industry, contributing to decarbonize the economy.

Paulo Mira Mourão

Department of Chemistry and Biochemistry, School of Science and Technology, University of Évora.

Paulo has a Physics and Chemistry graduation, a Master's in Physics, and a Ph.D. in Chemistry, Materials and Surface area. In the last years, he has consolidated his research in the fields of preparation, characterization, development, and application of porous materials, with the focus on the recovery and valorization of different raw materials and waste, from natural and/or synthetic sources, by its transformation into adsorbents with potential application in the liquid and gas phases, guided by a circular economy perspective. This knowledge and experience provide potential contributions in the areas of water treatment (e.g., drinking and wastewater), space management through the valorization of by-products and waste, of natural and synthetic origin, for use in the preparation of value-added materials (with multiple applications), among others (https://orcid.org/0000-0002-3634-2390).

Activated Carbon and Biochar: Complementary Carbon Materials

Abstract

Carbon materials are one of the research areas that have attracted significant interest from both academia and industry. This reflects increased knowledge transfer and collaboration between these two sectors, driven by the immense potential of these materials in a wide range of industries, including energy, chemicals, pharmaceuticals, food, healthcare, automotive, railways, aerospace, and, in particular, water treatment for both human consumption and wastewater management. Another advantage of carbon materials is that they can be produced from lignocellulosic and synthetic wastes, which are often problematic due to their nature and quantities. Thus, there is also the potential to valorize waste, especially natural and renewable waste, by converting it into high-value materials such as carbon-based adsorbents.

Among carbon-based materials, activated carbon and biochar stand out as versatile materials with complementary structural and chemical properties. Their diverse properties such as porous structure (including pore volume, average pore size, pore size distribution, and external surface area) and surface chemistry (specific chemical groups, point of zero charge) facilitate their application in water treatment.

This communication explores the potential of activated carbon and biochar as adsorbents. It presents and discusses the precursors used in their preparation, the production processes and their applications in both liquid and gas phases, supported by concrete examples developed at the University of Évora.

Round Table Event

Exploring agrivoltaics: Opportunities, trade-offs and potential solutions—integrating key aspects through a multiactor perspective discussion

Achieving carbon neutrality relies on renewable energy infrastructures like hydrological dams, photovoltaic arrays, and wind parks, often in agriculturally utilized areas. Despite their crucial role in achieving carbon neutrality while maintaining socio-economic growth, these infrastructures necessitate careful consideration of their impacts on land systems to achieve net emissions savings. Understanding the intricate interplay between renewable energy infrastructures, ecosystem functions, and political regulatory frameworks is essential for a successful transition to carbon neutrality. This round table discussion focuses on agrivoltaics—integrating perspectives from policy and academia to explore opportunities and trade-offs associated with photovoltaic infrastructures on agriculturally utilized land. By addressing technological, environmental, socio-economic, and political dimensions, the discussion aims to inform strategies for effective carbon emissions mitigation while respecting ecosystems and fostering a sustainable energy transition. It serves as a platform for interdisciplinary dialogue, promoting collaborative solutions for sustainable energy transitions.

Round Table Participants

Luís Fialho—Cátedra das Energias Renováveis, Universidade de Évora
Maximilian Stange—Fraunhofer IWU—Group Leader—Business Models for Sustainability
Filipe Alves—Biovilla, Cooperativa para o Desenvolvimento Sustentável e CHANGE, cEc3, Faculdade de Ciências de Lisboa
Miguel Sequeira—CHANGE, CENSE, Universidade Nova de Lisboa

Fernando Ascensão CHANGE, cE3c, Faculdade de Ciências de Lisboa
Representative from Public Administration

Round Table Moderator

Susana Filipe, CHANGE

Round Table Organizers

Dominik Noll
MED, CHANGE, Portugal

Flávio Oliveira
cE3c, CHANGE, Portugal

Susana Filipe
MED, CHANGE, Portugal

CHANGE
Global Change and Sustainability Institute

Contents

The Electric Vehicle Challenge for the Distribution Electricity Grid 1
José Gonçalves and Pedro Miguel

Sustainably Energy-Efficient Vegetable Dehydration Through Active Air Ventilation, Supported by Solar Thermal and PV Energy ... 13
João M. Garcia

Sludge from Household Setting Tank as a Raw Material Base for Obtaining Low Calorie Gaseous Fuel 21
Andrii Toporov and Olha Aleksieieva

Software-Based Fluid Dynamic Simulation of a Combustion Chamber 33
Alejandro Martínez-Martín, Alfonso Carlos Marcos Romero, Pedro José Muñoz Acebedo, and Juan Félix González González

Study of Corn Crop Residue Pelletising in a Semi-industrial System 43
R. García-Mateos, M. T. Miranda, J. I. Arranz, P. Romero, F. J. Sepúlveda, and S. Cuellar-Borrego

Assessing the Inclusiveness of the Draft Updated National Energy and Climate Plans 2030 of Belgium and Portugal Through a Hard-to-Reach Energy Users Lens 53
Miguel Macias Sequeira and João Pedro Gouveia

An Assessment of Wind Resources with Micro-Generation Applied on Farmland in Inland Portugal 65
Vanderlei Cruz, Natália Roque, António Canatário Duarte, and António Cerejo

Marginal Land Irrigated with Treated Wastewater. An Opportunity for Biofuel Production in Portugal? 77
Tânia Pires and Ana Maria Barreiros

Characterization of Refuse-Derived Fuel Pellets and Feasibility Assessment of Their Application in CO_2 Gasification 89
Santa Margarida Santos, Catarina Nobre, Margarida Gonçalves, and Paulo Brito

Hydrotreated Vegetable Oil (HVO): A Review on Production, Properties, Environmental Impact, and Future Perspectives 99
Fernanda Macedo de Araujo Azeredo, Cristina Moll Hüther, Julia Barbosa Werlang, and Roberta Panizio

Brewer's Spent Grain Valorization to Reduce the Energy Bill of Small Craft Breweries in Portugal 111
L. Rodrigues, C. M. Rangel, R. Panízio, G. Lourinho, and R. Traquete

Application of Renewable Energy—The Case of San Tomé Island 121
Ludcelma de Ceita, Ricardo Manso, and Teresa Eugénio

Torrefaction and Oxidative Carbonization of Pine Sawdust and *Chlorella Vulgaris* 141
Bruna Petry and Margarida Gonçalves

The Electric Vehicle Challenge for the Distribution Electricity Grid

José Gonçalves and Pedro Miguel

Abstract Electric vehicles (EV) in Portugal have increased from 2.455 units in 2015 to 80.465 units in 2021. In a context of uncertainty and market transformation, the International Energy Agency (IEA) estimate in its most favorable forecast, adapted to Portugal, a total of 428.044 EVs in 2030. This depicts an EV expansion of 4% in 2020, 17% in 2025, and 34% in 2030. However, such scenarios do not portray current market trends. The continued electrification of consumption result in an annual demand of 450GWh expected to occur between 2022 and 2027, representing an increment of 1%/year. Peak evolution is expected to increase 0.55%/year, from 8.41GW in 2021 to roughly 8.69GW in 2027. Using the registered consumption of primary substations (E-REDES—Distribuição de eletricidade. Open Data da E-REDES, 2022) as a capacity availability indicator show a 57% global peak demand index or roughly 19MVA in Portugal. A 40% available capacity was observed within considered primary substations. Despite the capacity of the network to accommodate the absolute energy demand for the transport sector, a significantly higher slope in EV evolution depicts a future challenge for the grid in a near future.

Keywords Electricity grid evolution · EV integration · Renewable disperse generation · Load management programs

1 Introduction

Transport is the only sector within the EU climate and energy policy framework that did not witness a reduction in greenhouse gas (GHG) emissions, over the course of the last thirty years [2]. The restrictions on GHG emissions until 2050, established by the European Green Deal, comprise a 90% reduction in all GHG transportation

J. Gonçalves (✉)
Innovation and Technological Standardization Department at E-REDES, Lisbon, Portugal
e-mail: joseantonio.goncalves@e-redes.pt

J. Gonçalves · P. Miguel
Institute for Systems Engineering and Computers at Coimbra, Coimbra, Portugal
e-mail: pcmiguel@gmail.com

P. S. Brito et al. (eds.), *Insights into Energy Trends*, Green Energy and Technology,
https://doi.org/10.1007/978-3-031-83811-8_1

emission compared to the 1990 emissions. Considering the path of increasing electrified consumption, the electricity grid must cope with more demanding and new challenges, in terms of infrastructure and its management. It no longer comprises an unidirectional flow of energy, with a growing number of prosumers, consumers that also produce electricity on site.

In Sect. 2 the characterization of the electricity distribution grid is addressed, in terms of capital investment, grid technologies, grid performance and forthcoming planned infrastructures meant to cope with new challenges. In Sect. 3 the current and the expected future deployment of electric vehicles is exhibited, its features and accompanying infrastructure. The main drive for the boost in this technology is to have lower road GHG associated emissions, due to decreasing GHG emission intensity of electricity generation. Conclusions of this work are presented in Sect. 4.

2 The Electricity Distribution Grid Characterization

E-REDES is the main Portuguese distribution system operator (DSO) maintaining nearly 84 Mm of HV and MV electric lines. LV electric lines account for 147 Mm and are feeded by more than 430 primary substations and 70 k secondary substations [3] in mainland Portugal.

2.1 The Electricity Distribution Grid Characterization

Responsible for maintaining "the lights on" for roughly 6.4 M clients, spread over 278 municipalities [4]. In 2021, E-REDES had a total annual investment of 348 M€ and a maintenance annual cost of nearly 117 M€. E-REDES is also seeking to increase the quality of technical service, currently with 50 min. in the TIEPI MV, 80 min. in the SAIDI LV and 1.57 in SAIFI LV [3]. Considering the reasonable investment amount to maintain the electricity grid and the aim to maximize results, the DSO (Distribution System Operator) has to standardize as much as possible the technological solutions to be implemented.

The DSO investment strategy considers several drivers such as: the technical and financial requisites, the energy policy guidelines, described in the most recent Security of Supply Monitoring Report with horizon 2021–2040 (RMSA-E 2020) [5], the energy transition objectives and goals of the National Energy and Climate Plan 2021–2030 [6], as well as, all legal and social-economic government recommendations [7].

From the technical side, challenges such as the significant integration of prosumers, disperse generation, energy communities and the transport electrification need to be addressed, while addressing the economic national and worldwide challenges. In [7] two approaches are presented to predict the demand evolution until 2027. The medium scenario projection, used in the study [7] considers a non-exceeding demand of 50% (in the same study the lower scenario considered 33% and the upper scenario considered 66% of non-exceeding demand).

Considering the medium scenario results in Fig. 1 (yellow continuous line) an annual demand evolution of nearly 450 GWh is expected to occur between 2022 and 2027, representing an average increment of roughly 1%/year.

In terms of peak evolution, studies point to an increase of nearly 0.55%/year in the required capacity of the grid, as presented in Fig. 2, evolving from 8.41GW in 2021 to roughly 8.69 GW in 2027 in medium scenario [7].

In Fig. 2 it can be depicted the influence of the national financial crisis occurred between 2011 and 2015. After this period the consumption start to modestly augment and more recent projections for the 2021 to 2027 timeframe is to augment nearly

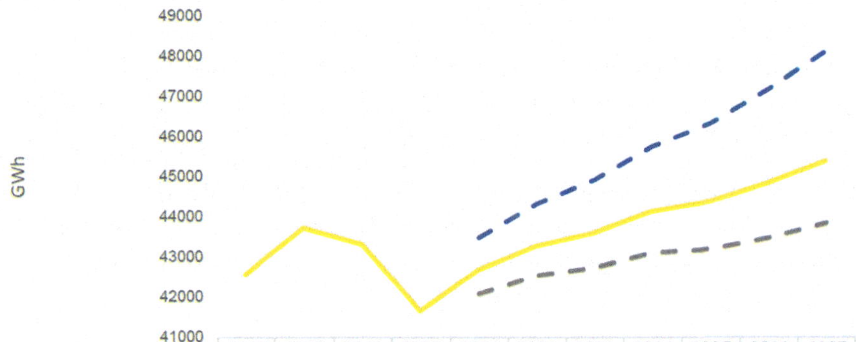

	2017	2018	2019	2020	2021	2022	2023	2024	2025	2026	2027
Upper Scenario					43 47	44 31	44 90	45 74	46 31	47 17	48 11
Central Scenario	42 59	43 75	43 34	41 68	42 70	43 30	43 63	44 16	44 40	44 87	45 41
Lower Scenario					42 06	42 52	42 72	43 09	43 18	43 46	43 83

Fig. 1 Required energy for distribution grid, without HV demand [7]

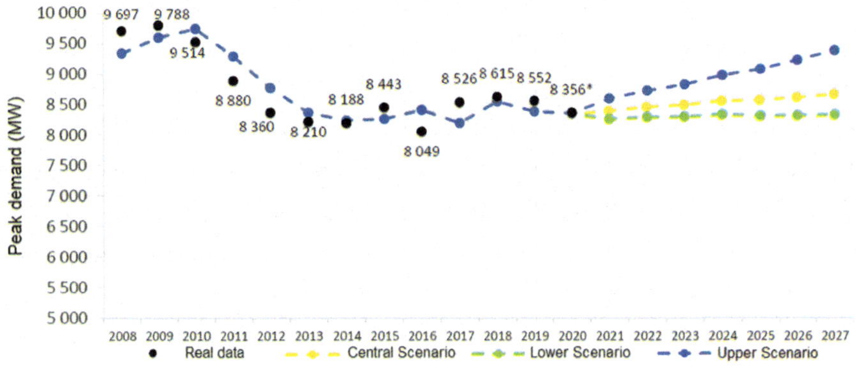

Fig. 2 Capacity evolution predictions of the Portuguese main DSO [7]

3.86%, meaning 8.685GW peak in the consumption diagram, in the medium scenario [6].

2.2 Grid Technologies

The DSO primary substation state of the art is maintained by the E-REDES standardization and innovation department [8] and is, since 2007 characterized in three major primary substation type projects, namely the simplified primary substation (DIT-C13-504), the external base primary substation (DIT-C13-503) and the external primary substation (DIT-C13-500) [9]. The previous type projects differentiate by the following characteristics, presented in Table 1.

In terms of secondary substation, the Portuguese DSO have both poll mounted and ground mounted solutions for rural and urban areas, respectively. The rated characteristics of those standardized solutions are summarized in Table 2.

The 1000kVA power transformers capacity is a recently standardized solution (since 2022) for secondary substations to address special cases in urban areas with high power requirements and extremely low implantation area. Before this recent update the technical solutions were mainly supported in capacities of 400 and 630kVA. All the previous solutions are standardized for the distribution grid in Portugal considering the rated voltages of 10 kV, 15 kV and 30 kV, and are presently with an average capacity availability between 33 and 50% [1].

Table 1 Primary substation standardized types

	Exterior primary substation (DIT-C13-500)	Exterior base primary substation (DIT-C13-503)	Simplified primary substation (DIT-C13-504)
Application:	Rural or semi-urban areas;	Rural areas	Rural or semi-urban areas
Rated voltage ratio:	60/30, 60/15 ou 60/10 kV		
Rated Power capacity:	2 × 40 MVA	1 × 40 MVA	1 × 31.5 MVA
HV panels:	6 LHV + 2 PT	3 LHV + 1 PT	1 Panel (Line/transformer);
MV panels:	12 bays in 30 kV	6 bays in 30 kV	4 bays in 30 kV
	20 bays in 15 kV	10 bays in 15 kV	6 bays in 15 kV
	or 24 bays in 10 kV	or 12 bays in 10 kV	
Power buses:	2 buses in HV and 2 buses in MV side	1 bus in HV and 1 bus MV side	One MV bus

Table 2 Secondary substation standardized capacities

Application type		Rated capacity
		S [kVA]
Poll mounted	R100	50
		100
	R250	160
		250
Ground mounted		400
		630
		800
		1000

2.3 *The Grid Performance*

While considering technological and financial boundaries, the Portuguese grid indicators had evolved in a highly impactful path towards excellency. As depicted in Fig. 3, the quality of service (QoS) indicators have improved nearly 90% in the last 20 years, both for the time of equivalent interruptions per power installed (TIEPI) and system average interruption duration index (SAIDI) [10, 11].

Using registered consumption for the primary substations [1], a 57% global peak demand index for capacity availability can be observed, higher than the ideal reference of 50% and representing roughly 19 MVA. This indicator is on average segregated as 7.3 MVA for 20 MVA primary substations, 13.6 MVA for primary substation with capacity between 20 and 31.5 MVA, 16.6 MVA peak demand for primary substation with capacity between 31.5 and 40 MVA and 30.4 MVA peak demand for primary substation with capacity above 40 MVA. In this assessment standardized and non-standardized primary substations were considered, of which roughly 40% capacity availability was observed.

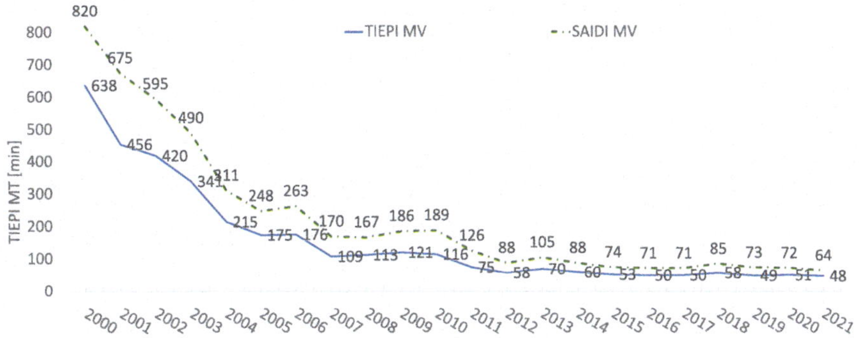

Fig. 3 Evolution of quality of service indicators [11]

3 The Electric Vehicle Evolution

According to [12], the electric car market is expected to grow [8] worldwide. Over 16.5 million electric vehicles (EV) were on roads in 2021 and are its numbers are expected to triple in just three years. The increasing deployment of EV is driven by continued policy as its main pillar. Such policies include public spending on subsidies and incentives for EV, the phase out of internal combustion engines, ambitious vehicle electrification targets and the incentive to public transportation.

The interest boost within the car market stimulates manufactures to provide consumers with increased choice of EV, electrifying their fleets, and one key opportunity to control emissions of GHG (Green House Gases) [2].

In Fig. 4, greenhouse gas emission intensity (gCO_{2e}/kWh) is presented, calculated as the ratio of CO_{2e} emissions from public electricity production, as a share of CO_2 equivalent emissions from public electricity and heat production related to electricity production, and gross electricity production.

The 2030 values are predictive intensity levels that would allow the EU to achieve a net 55% reduction in greenhouse gases by 2030, compared with 1990. An incentive to the electrification of consumption in observed as the GHG decreases in the intensity of electricity generation.

3.1 EV Stock Increase in Portugal

The EV stock has increased significantly from the 2,455 units in 2015 to the 50,897 in 2020 or even to 80,465 units in 2021. Nearly 39,142 are BEV and 41,323 are PHEV.

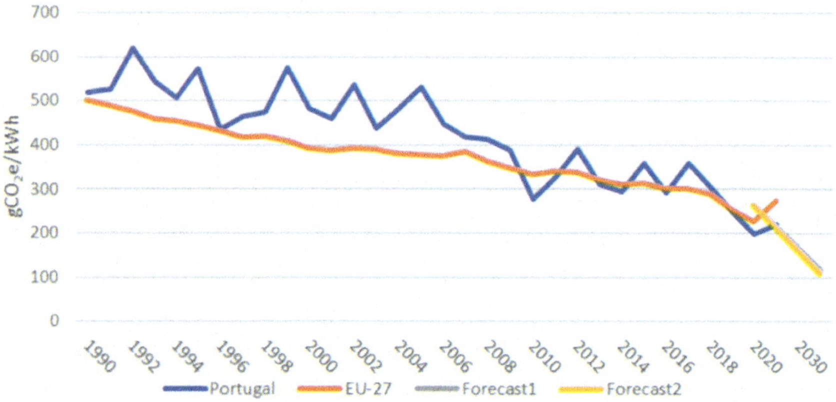

Fig. 4 Greenhouse gas emission intensity of electricity generation [2, 13]

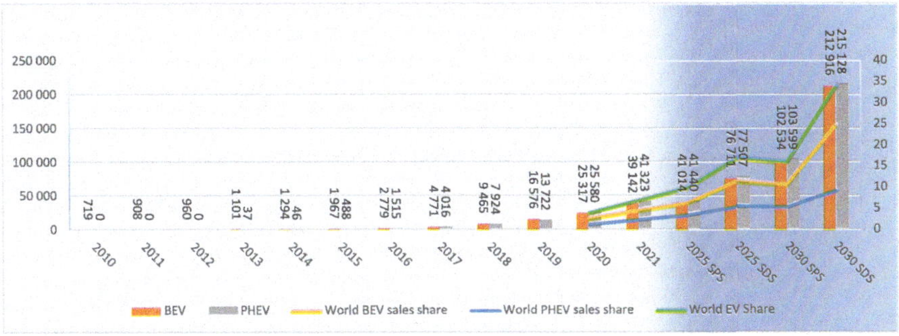

Fig. 5 EV stock in Portugal [12]

Applicable to the European and Portuguese market, two possible scenarios were developed by IEA to forecast sales share and the expected number of EV over time.

The Stated Policies Scenario (SPS) reflects existing policies and measures, as well as policy ambitions and targets that have been legislated by governments around the world.

The Sustainable Development Scenario (SDS) assumes that the announced ambitions and targets made by governments around the world, including the most recent ones, are met in full and on time. With regards to electromobility all recent major announcements of electrification targets and longer-term net zero emissions are included as well as other pledges, regardless of whether these have been anchored in legislation or in updated Nationally.

The most conservative forecast SPS, adapted to Portugal, predicts in 2030, a total number of 206,133 EVs, of which 102,534 are BEV and 103,599 are PHEV. Such analysis, as depicted in Fig. 5 combines a shared world sale, supported in real values until 2020 and in predicted forecasted for the following years, of 4%, 10% and 16%, for 2020, 2025 SPS and 2030 SPS, between BEV and PHEV.

On other side a more radical approach presents the SDS scenario forecasts with an EV expansion of 4% in 2020, to 17% in 2025, to 34% in 2030. Such scenarios still consider small deployment of electric vans and buses, while trucks are not even considered.

3.2 *The Capacity Evolution*

The increasing number of EVs and the expected shift in the resources used to power vehicles represents a challenge to the electricity grid. As perceivable from the data in Fig. 6, the number of public EV charging stations almost quadruplicated to 4,178, between slow and fast charging points with a trend to an exponential growth.

In 2016, when EVs where not so trendy, the Statista Research Department estimated average battery capacity in electric vehicles worldwide from 2017 to 2025,

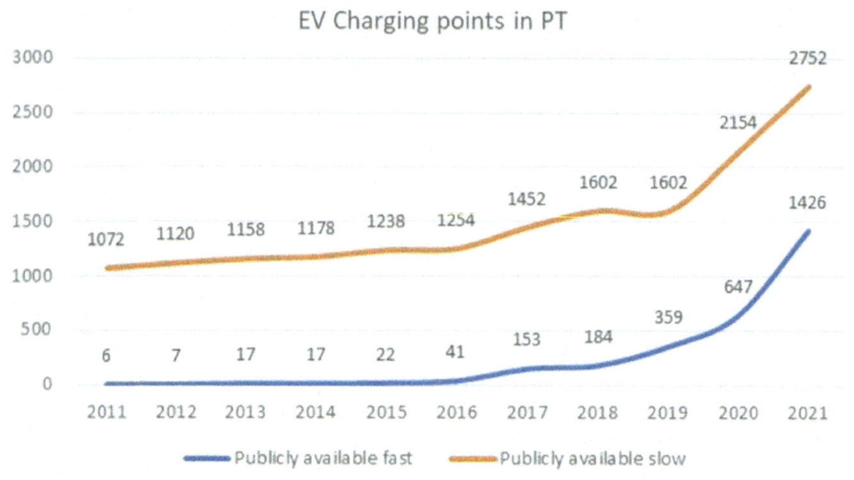

Fig. 6 Charging points in Portugal [12]

by type of vehicle. In a conservative perspective, and considering that in 2030, the average battery capacity is the same as the one predicted to 2025, it was calculated the total battery capacity of EV's in Portugal, from 2017 to 2030.

Considering an average battery capacity in PHEV and BEV of 10.2 kWh and 41 kWh before 2017 and of 10.3 kWh and 43 kWh until 2021, the evolution rate, presented in Fig. 7 can be estimated.

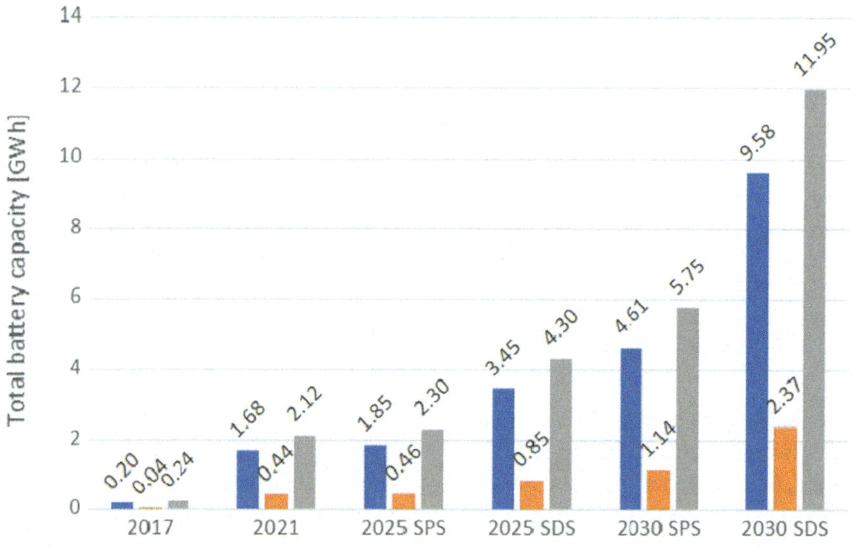

Fig. 7 Total battery capacity of EVs in Portugal [12]

Obtained results enable us to verify an average EV battery capacity annual evolution of more than 150% for the last 10 years.

As expected the most conservative scenarios (SPS), when compared to 2020 projections, exhibits an evolution rate of 169% and 423%, in 2025 and 2030, respectively. The sustainable development scenario (SDS) present an evolution rate of 317% and 879%, in 2025 and 2030, also in the increase of battery capacity.

In Table 3 the annual evolution rate of battery capacity for the transport infrastructure, considering the SPS and SDS projections, are estimated between 42 and 79%, respectively, considering the 2021–2025 time spam. And could vary between 42 and 88%, considering the SPD and the SDS scenarios between 2025 and 2030, respectively.

The increase in BEV and PHEV capacity presents a higher slope than the demand and grid capacity slope evolution, as described in Sect. 2.1. This fact becomes more relevant as the number of EV and the number of charges/EV increases (Table 4).

Table 3 Capacity availability (CA) in primary substations [7]

	20 MVA PS		[20; 31.5] MVA PS		[31.5; 40] MVA PS		Higher than 40 MVA PS	
	CA [MVA]	LI [%]	CA [MVA]	LI [%]	CA [MVA]	LI [%]	CA [MVA]	LI [%]
	7.29	58%	13.60	56%	16.56	58%	30.40	55%
N.º of facilities	94		48		104		148	
General capacity availability [MVA]								19.23
Global load index [%]								56.6%

CA—Capacity availability, PS—Primary substations, LI—Load index

Table 4 EV capacity evolution estimation [12]

Year	Number of BEV	Number of PHEV	Battery capacity for BEV [GWh]	Battery capacity for PHEV [GWh]	Estimated Energy for EV [GWh]	Evolution rate—battery capacity (%)
2010	719	0	0.03	0.00	0.03	
2011	908	0	0.04	0.00	0.04	126
2012	960	0	0.04	0.00	0.04	106
2013	1 101	37	0.05	0.00	0.05	116
2014	1 294	46	0.05	0.00	0.05	118
2015	1 967	488	0.08	0.00	0.09	160
2016	2 779	1 515	0.11	0.02	0.13	151
2017	4 771	4 016	0.20	0.04	0.24	183
2018	9 465	7 924	0.41	0.08	0.49	208
2019	16 576	13 722	0.71	0.15	0.86	175
2020	25 317	25 580	1.09	0.27	1.36	158
2021	39 142	41 323	1.68	0.44	2.12	156
2025 SPS	41 014	41 440	1.85	0.46	2.30	**169**
2025 SDS	76 711	77 507	3.45	0.85	4.30	**317**
2030 SPS	102 534	103 599	4.61	1.14	5.75	**423**
2030 SDS	212 916	215 128	9.58	2.37	11.95	**879**

4 Conclusion

The present paper presents the distribution grid characterization and its implications in terms of capital investments and new present challenges.

The massive integration of prosumers, disperse generation, energy communities and the transport electrification are some of the accountable variables to predict the demand evolution and the required network investments.

Different studies try to estimate the evolution demand associated to transport electrification. The authors present in the paper possible approaches to estimate the EV energy demand, however, the unpredictability of user patterns, the different battery capacities and the battery technological evolutions are variables very difficult to predict and with constant automotive market variations, especially because they are also influenced by socio-economic and geopolitical aspects.

Despite the annual demand evolution medium projection of roughly 1%/year (i.e. near 450 GWh), and the peak evolution of nearly 0.55%/year (i.e. roughly 8.69GW in 2027) a peak demand index of nearly 57% is presently observed in the primary substations.

In terms of EV evolution, and the objective to reduce the transport greenhouse gases by 55%, the EV stock in increasing significantly, and therefore presenting an important challenge to be addressed. The obtained results enable us to verify the average annual EV battery capacity evolution supported in data from the last 10 years.

Despite the present capacity of the network to accommodate the associated absolute energy demand for the transport sector, the significantly higher slope of the EV evolution depicts a near future challenge for the grid.

If projections regarding continuous electrification materialize, even in the SPS scenario, with 169% and 423% battery capacity growth, as shown in Fig. 5 and Table 3, in 2028 and 2040, and if no additional measures are taken to improve grid resilience, a critical point is expected to be reached within 4–12 years. With such growth rates, the future EV battery capacity is equaled to current energy flow at primary substations.

The need to ensure improved grid stability indicators and service along with present and new challenges requires a permanent observation and analysis of the electricity distribution grid capacity, in the present and in the future.

Funding This work has been supported by the Portuguese Foundation for Science and Technology (FCT) under project grant UIDB/00308/2020.

References

1. E-REDES—Distribuição de eletricidade (2022) Open Data da E-REDES
2. European Environment Agency Transport and environment report 2021—Decarbonising road transport—the role of vehicles, fuels and transport demand. https://doi.org/10.2800/68902
3. E-REDES—Distribuição de eletricidade (2021) EREDES em números 2021
4. E-REDES—Distribuição de eletricidade A quem se destina. https://www.e-redes.pt/pt-pt/quem-se-destina. Accessed 14 Dec 2022
5. DGEG Relatório de Monitorização de Segurança de Abastecimento do Sistema Elétrico Nacional, para o período de 2021–2040 (RMSA-E 2020)
6. Portuguese government Plano Nacional Energia e Clima 2021–2030 (PNEC2030)
7. E-REDES PDIRD-2020—Plano de Desenvolvimento e Investimento 2021–2025
8. E-REDES—Distribuição de eletricidade Organograma E-REDES
9. E-REDES—Distribuição de eletricidade Catalogo de documentos normativos. https://www.e-redes.pt/sites/eredes/files/2022-11/DLN-Z40-010%20Ed3_2022.pdf. Accessed 14 Dec 2022
10. Gonçalves X, Miguel P, Neves L, Ramalheira F, Rosa J, Santos BO (2021) The past, present and future role of automation in the electricity grid—A Portuguese perspective. In: CIRED 2021—The 26th international conference and exhibition on electricity distribution. Institution of Engineering and Technology, pp 1617–1621
11. E-REDES—Distribuição de energia (2021) Relatório de Qualidade de Serviço 2021
12. Energy Agency I Global EV Outlook 2020 Entering the decade of electric drive?
13. European Environmental Agency (2020) Greenhouse gas emission intensity of electricity generation in Europe. https://www.eea.europa.eu/data-and-maps/indicators/overview-of-the-electricity-production-3/assessment. Accessed 27 Dec 2022

Sustainably Energy-Efficient Vegetable Dehydration Through Active Air Ventilation, Supported by Solar Thermal and PV Energy

João M. Garcia

Abstract This study introduces an energy-efficient and cost-effective solar vegetable dehydrator that integrates air ventilation, solar thermal, and photovoltaic energy for operation in various weather conditions. In its prototype phase, the dehydrator aims to process small farm product quantities, reducing waste by dehydrating unsalable fruits for marketing as dried fruit. This promotes sustainability among small-scale agricultural producers. Optimization focuses on processing within a suitable timeframe, ensuring high energy efficiency, low costs, and enhanced product quality with desirable flavor and aroma. The dehydrator's active control of supply air and regulation of air circulation velocity are achieved through real-time monitoring of ambient temperature and humidity at the inlet and outlet. This dynamic control allows adjustments in operating conditions by managing the mix of fresh and recirculated air flows. Designed for mobility, the device facilitates easy relocation for optimal dehydration, enhancing operational efficiency and adaptability to local conditions. Through the developing and implementing this equipment, the study showcases the potential use of damaged fruits for commercial purposes. Successfully selling thinly sliced and dehydrated products contributes to the economic valorization of agricultural activities and promotes overall sustainability.

Keywords Sustainability · Solar energy · Vegetable dehydration · Air ventilation

J. M. Garcia (✉)
Lisbon Superior Institute of Engineering, Lisbon, Portugal
e-mail: joao.garcia@isel.pt

UnIRE, ISEL, Polytechnic University of Lisbon, Lisbon, Portugal

MARE-IPS, Marine and Environmental Sciences Centre, Escola Superior de Tecnologia, Instituto Politécnico de Setúbal, 2910-761 Setúbal, Portugal

P. S. Brito et al. (eds.), *Insights into Energy Trends*, Green Energy and Technology,
https://doi.org/10.1007/978-3-031-83811-8_2

1 Introduction

Dehydration stands out as a straightforward and economical technique for preserving food, involving the extraction of adequate moisture to facilitate prolonged preservation. The success of food dehydration hinges on the meticulous control of three pivotal operational parameters: temperature, relative humidity, and air velocity [1]. Precision in regulating temperature is crucial to prevent the risk of over-drying while effectively removing moisture from the food [2]. Maintaining a low relative air humidity is indispensable for optimizing the absorption of released water from the food.

Ensuring an appropriate fresh airflow rate is essential for expeditiously removing water vapour from the dehydration chamber [3]. The specific aspect of vegetable dehydration follows the same principles. Various methods, including solar radiation, an oven, or an electric dehydrator, can be utilized for vegetable dehydration, with a primary emphasis on maintaining the optimal temperature, humidity, and air velocity tailored to the specific type of food [4]. Vegetable dehydration temperature ranges vary distinctly based on horticultural products, ranging from 35 °C for herbs to 57 °C for fruits [5]. Although air velocity is not the sole determinant in the dehydration process, several studies have indicated that speeds of up to 1.5 m/s are suitable for this procedure [6]. However, it is noteworthy that the moisture content and temperature of the air within the dehydration chamber exhibit a direct correlation with the airflow rate [7]. Thus, possessing the capability to regulate airflow rate stands as a significant advantage in such scenarios.

2 Dehydration Principles

2.1 Conception

The conceptualized equipment constitutes a solar dehydrator for horticultural purposes, harnessing both solar thermal and photovoltaic energy resources. This modular, cost-efficient, and portable system represents an energy-conscious, economical, and flexible solution, specifically engineered to function entirely on renewable energy sources—solar thermal and photovoltaic—ensuring independence from weather variations [8, 9]. However, in exceptional situations, it can adapt to conventional electric power [10]. As previously stated, the efficacy of the dehydration process relies on the meticulous control of three pivotal factors: Temperature (T), Relative Humidity (RH), and Air Velocity (V) [11].

2.2 Constituent Parts

The dehydrator comprises modular elements, encompassing a solar thermal panel, a photovoltaic solar panel, a dehydration chamber, an electric resistance, air circulation fans, motorized air damper actuators/valves, temperature and humidity probes, and a control system guaranteeing the operation of the system under defined conditions. Figure 1 illustrates the operational principle, highlighting all components.

Following a brief description of the functions of the main components:

The Solar Thermal Panel, custom-built for heating the dehydration chamber air under suitable weather conditions, plays a pivotal role in maintaining the desired temperature during the drying process. Simultaneously, the Photovoltaic Panel (PV) takes charge of charging batteries that power critical support devices, ensuring continuous operation of the dehydrator's system components, including the controller, air fans, electric resistance, and motorized actuators/valves. Within the Dehydration Chamber, multiple shelves facilitate the arrangement of products for dehydration. Dry and heated air, orchestrated by the PV-fed electric resistance, circulates through these shelves, optimizing the dehydration process for fruits and vegetables. Acting as a backup heating source, the PV-Fed Electric Resistance activates when the air temperature falls below the set point, ensuring consistent heating during adverse weather conditions. Variable-speed Air Fans, integrated into the control system, regulate air circulation, adapting speed to achieve efficient drying. The Mixing Module, equipped with adjustable registers, enhances control over the composition of fresh air in the dehydrator chamber's supply airflow. Meanwhile, Temperature and Relative Humidity Probes continuously monitor conditions, allowing real-time adjustments for optimal dehydration. Serving as the dehydrator's "brain," the Control Module

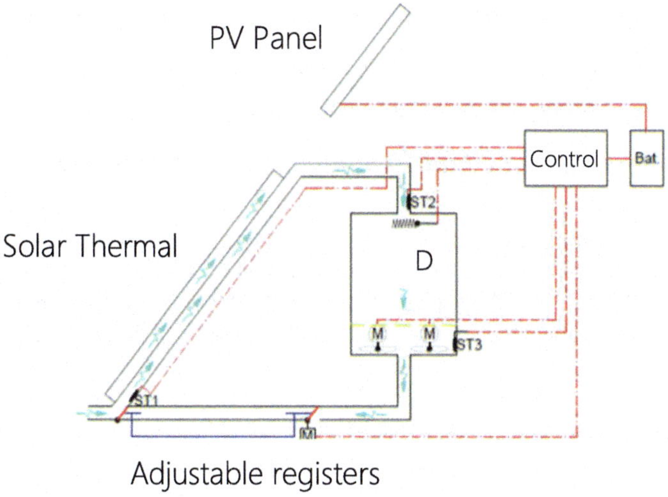

Fig. 1 Basics scheme

oversees air inlet and outlet, temperature, relative humidity, and air velocity. It ensures a seamless and controlled product dehydration process by managing these critical parameters.

2.3 Process Basics

In order to maximize the efficiency of the drying process, precise control over various variables is imperative, including air temperature, relative humidity, air circulation speed within the apparatus, and air registers [12]. This control is orchestrated by the system controller. The drying methodology employed by this device involves directing heated air into the dehydration chamber, primarily sourced from solar energy, resulting in substantial energy conservation [13]. Optimal equipment performance necessitates specific climatic conditions, such as elevated temperatures (>60 °C), moderate air velocity, and low relative humidity [14]. The process of food dehydration involves placing products inside a sealed chamber equipped with multiple shelves to facilitate proper air circulation. These conditions help maintain low relative humidity, ensuring effective dehydration [15]. This dehydrator stands out as a hybrid system, proficient in utilizing exclusively renewable energy or, in less favorable weather conditions, conventional electricity. This dual capability ensures the provision of necessary conditions for the dehydration process. An innovative feature of this equipment lies in its adaptability to either fresh air circulation or recirculated air, contingent on current weather conditions. This flexibility is achieved through the strategic positioning of actuators/motorized valves, enhancing the functionality of the dehydrator. Moreover, variable-speed fans contribute to achieving the ideal air velocity for dehydration, aligning with user-defined objectives-whether expediting the process for time optimization or slowing it down for enhanced taste and odor in the dehydrated product.

2.4 Computational Fluid Dynamic Simulations (CFD)

To study the optimizations of the drying process, the dehydration chamber was simulated using a CFD numerical simulation model. ANSYS Fluent software was employed to simulate and optimize the optimal airflow inside the drying chamber. This numerical model solves the mass and momentum transport equations for constant density, with the viscous dissipation term considered negligible [16]. Two turbulence equation models determine a turbulent length and time scale by solving two independent transport equations. The simulated geometry was the dehydration chamber with six shelves made of food-grade netting. However, since the produce slices were placed tightly on the shelves, obstructing the airflow, the shelves were considered solid obstacles in the simulations. The model used a hybrid, structured mesh with approximately 2 × 10^6 elements and was run on an Intel Xeon W-2155

3.30 GHz processor with 63 GB RAM and an NVIDIA Quadro P4000 GPU [17]. The SIMPLE algorithm was utilized to couple velocity and pressure fields. The equations for turbulent quantities and momentum were discretized using second-order upwind schemes, while the PRESTO scheme was used for pressure. The convergence criteria for the mass, velocity, and turbulent quantities transport equations were set to 10^−3. The numerical study aimed to map the air distribution inside the chamber to determine if and where different shelves required more or less airflow. The boundary conditions were set as follows: Inlet velocity was 0.1 m/s and 2.9 m/s for different scenarios; and Outlet conditions were set as Outflow. Temperature was not considered in the simulations, as it is not relevant to the air distribution within the chamber, which depends on air velocity and chamber geometry. Additionally, temperature can be influenced by atmospheric conditions and varies depending on the specific dehydration temperature of the product being dried. CFD simulation work helps understand and demonstrate the importance of airflow within the dehydration chamber and allows for the optimization of the final design.

2.5 *Operating Principle*

The proposed dehydrator operates under three distinct conditions: using fresh air, recirculated air, or a combination of both. The 100% fresh air condition is activated when ideal external parameters, such as high temperatures, moderate wind, and low relative humidity, typically found during summer or in hot and dry weather, are met. Under these conditions, outside air enters through the air inlet located at the base of the thermal solar panel. It rises due to natural convection, driven by the temperature difference resulting from solar radiation absorption, until it reaches the entrance of the dehydration chamber. Inside the chamber, air is forced downward by natural convection and aided by air blowers located in the lower part of the chamber, directing the air outward. For this condition, air registers are positioned to allow air intake at the base of the thermal solar panel and air expulsion downstream of the air circulation fans. These registers are activated by an actuator controlled by the system controller, ensuring that the air inside the dehydration chamber is constantly renewed.

When outside parameters are not ideal for using 100% fresh air, typically in winter or during cold, cloudy, and relatively dry weather conditions, the air is recirculated within the chamber and the thermal solar panel. The air circulates through the equipment, entering from the base of the thermal solar panel to the entrance of the dehydration chamber due to natural convection driven by the temperature difference resulting from solar radiation absorption. Fans located in the lower part of the dehydration chamber then force the air downward, and it returns to the entrance of the thermal solar panel. In this scenario, the air registers are closed, restricting the intake of outside air and the outlet of inside air, until sensors detect that the circulating air is saturated with humidity and needs to be renewed with drier, albeit colder, outside air. If the thermal solar panel alone cannot achieve the ideal temperature inside the chamber, a thermal resistor can provide additional assistance.

The mixed air mode is designed for weather conditions that are neither cloudy nor humid. In this mode, a portion of the circulating air is reused and mixed with fresh air inside the dehydration chamber. This blend of recirculated and fresh air rises through the thermal solar panel due to the temperature difference and natural convection, driven by solar radiation warming the panel (and possibly the resistor). It reaches the inlet of the dehydration chamber and is then directed downward by fans located in the lower zone of the chamber. To replace some of the circulating air with fresh air, the air register at the bottom is activated, while the remaining air is directed toward the inlet of the thermal solar panel for mixing with fresh air. The control module regulates the dampers based on temperature and humidity sensor measurements to manage the inlet and outlet of air.

2.6 Testing and Outcomes

Following the completion of the prototype construction, a series of comprehensive tests were undertaken on the apparatus, involving the dehydration of various horticultural products. The dehydrated horticultural items consistently demonstrated remarkable taste and texture, accompanied by notably reduced dehydration durations in comparison to equivalent processes carried out using conventional electric equipment. Figure 2 illustrates the progression of a performance test, while Table 1 showcases specific results for dehydrated eggplants.

Numerous performance tests were meticulously designed and executed, encompassing a range of vegetable types and diverse atmospheric conditions such as solar radiation, air temperature, and air velocity. Table 1 provides an illustrative example of the results obtained from these tests, while specific data regarding the initial and final moisture content for eggplants are detailed in Table 2.

Fig. 2 Real tests and prototype analysis

Table 1 Tests for eggplants dehydration process

Hour	T1 (°C)	T2 (°C)	T3 (°C)	T4 (°C)	H3 (%)	H4 (%)	Resistance	Type of air
11:00	31.3	60.3	49.4	35.7	–	–	On	Recirculation
11:35	35.6	60.4	52.5	39.9	–	–	Off	Exterior
12:15	30.9	62.6	52.8	43.4	–	–	Off	Exterior
12:30	33.5	62.3	50.8	42.1	–	–	Off	Exterior
13:00	31.4	62.8	49.5	36.8	–	–	Off	Exterior
13:30	32.6	62.2	50.1	37.7	12.3	33.5	Off	Exterior
14:00	32.2	64.6	51.6	37.9	10.1	30	Off	Exterior
14:30	34.6	62.4	52.7	38.6	9.1	28.7	Off	Exterior
15:00	37.6	62.9	52.7	38.6	9.1	28.7	Off	Exterior
16:00	39.6	63.7	52.7	41.8	8.9	22.6	Off	Exterior
16:40	41.7	64.2	54.8	43.6	7	19.5	Off	Exterior
17:00	41.7	65	55	43.5	5	16.9	Off	Exterior
17:15	38.4	63.7	55.3	43	5.4	17.1	Off	Exterior
17:30	37.2	62	54.7	42.2	5.4	17.9	Off	Exterior

Table 2 Initial and final moisture content results (Eggplants)

Moisture content (%)	Test 1	Test 2	Test 3	Test 4	Test 5
Fresh	92.3	91.8	93.1	91.9	92.1
Dried	2.1	1.9	2.2	1.7	1.7

3 Conclusions

A solar-powered horticultural dehydrator has been elucidated in this study. This dehydrator, distinguished by its energy efficiency, cost-effectiveness, modularity, and portability, relies exclusively on renewable energy sources—solar thermal and photovoltaic. This design allows it to operate autonomously, irrespective of prevailing weather conditions.

To enhance the dehydration process, crucial factors including time, product quality, energy efficiency, cost, and flavor/odor were meticulously addressed in the development of an efficient dehydrator. Optimization of these factors was achieved through precise control of dehydration air temperature and air circulation velocity, with adjustments made based on the specific product being dehydrated, prevailing solar radiation, and external temperature conditions.

The innovative dehydrator actively utilizes real-time outdoor air conditions, continuously optimizing the process by monitoring temperature and relative humidity levels of external air, air entering the dehydration chamber, and air exiting the chamber. This optimization is facilitated through a recirculated air and airflow control system, ensuring the equipment operates under optimal conditions consistently.

Performance tests conducted on the developed system have yielded the conclusion that the dehydration process has been successfully optimized. This optimization is attributed primarily to the innovative design configuration, particularly the dehydration chamber, which maximizes the efficiency of the dehydration process.

References

1. Noor Mohammed A, Chauhan OP, Semwal AD (2024) Emerging technologies for fruits and vegetables dehydration, Food Human 2:100303. https://doi.org/10.1016/j.foohum.2024.100303
2. Akter J et al (2024) Colour, nutritional composition and antioxidant properties of dehydrated carrot (Daucus carota var. sativus) using solar drying techniques and pretreatments. Heliyon 10(2):e24165. https://doi.org/10.1016/j.heliyon.2024.e24165
3. Wang B et al (2023) Dehydration–rehydration vegetables: evaluation and future challenges. Food Chem X 20:100935. https://doi.org/10.1016/j.fochx.2023.100935
4. Gilago MC, Mugi VR, CVP (2023) Evaluation of drying kinetics of carrot and thermal characteristics of natural and forced convection indirect solar dryer. Results Eng 18:101196. https://doi.org/10.1016/j.rineng.2023.101196
5. Mohammed SA, Alawee WH, Chaichan MT, Abdul-Zahra AS, Fayad MA, Aljuwaya TM (2024) Optimized solar food dryer with varied air heater designs, Case Stud Thermal Eng 53:103961. https://doi.org/10.1016/j.csite.2023.103961
6. Madhankumar S, Viswanathan K, Taipabu MI, Wu W (2023) A review on the latest developments in solar dryer technologies for food drying process. Sustain Energy Technol Assess 58:103298. https://doi.org/10.1016/j.seta.2023.103298
7. Riaz S, Kabir A, Haroon A, Ali A, Faisal Manzoor M (2023) Food dehydration recent advances and approaches, a comprehensive review of the versatile dehydration processes. IntechOpen. https://doi.org/10.5772/intechopen.108649
8. Wang X et al (2020) A systematic review of recent air source heat pump (ASHP) systems assisted by solar thermal, photovoltaic and photovoltaic/thermal sources. Renew Energy 146:2472–2487. https://doi.org/10.1016/j.renene.2019.08.096
9. Geetha P et al (2023) Smart operating range monitoring of solar PV cell with integrated phase change materials by using solar deep learning model. Electric Power Compon Syst 1–12. https://doi.org/10.1080/15325008.2023.2249882
10. Nagavani C, Deepika T, Maria Ancy Shinesha W, Matcha Soundarya M (2024) Dehydration of food materials using solar dryer with mobile app integration. J Electr Eng Autom 6(1):72–81. https://doi.org/10.36548/jeea.2024.1.006
11. Li Y et al (2024) Factors affecting chemical and textural properties of dried tuber, fruit and vegetable. J Food Eng 365:111828. https://doi.org/10.1016/j.jfoodeng.2023.111828
12. Suresh BV, Shireesha Y, Kishore TS, Dwivedi G, Haghighi AT, Patro ER (2023) Natural energy materials and storage systems for solar dryers: state of the art. Solar Energy Mater Solar Cells 255:112276. https://doi.org/10.1016/j.solmat.2023.112276
13. Catorze C et al (2022) Study of a solar energy drying system—Energy savings and effect in dried food quality. Energy Reports 8:392–398. https://doi.org/10.1016/j.egyr.2022.01.070
14. Chavan A, Vitankar V, Mujumdar A, Thorat B (2021) Natural convection and direct type (NCDT) solar dryers: a review. Dry Technol 39(13):1969–1990. https://doi.org/10.1080/07373937.2020.1753065
15. Pateiro M et al (2022) The role of emerging technologies in the dehydration of berries: quality, bioactive compounds, and shelf life. Food Chem X 16:100465. https://doi.org/10.1016/j.fochx.2022.100465
16. Alvarenga N, Martins J, Caeiro J, Garcia J, Pássaro J, Coelho L, Santos MT, Lampreia C, Martins A, Dias J (2021) Applying computational fluid dynamics in the development of smart ripening rooms for traditional cheeses. Foods 10:1716. https://doi.org/10.3390/foods10081716
17. Garcia JM, Rosa A (2019) Theoretical study of an intermittent water-ammonia absorption solar system for small power ice production. Sustainability 11:3346. https://doi.org/10.3390/su11123346

Sludge from Household Setting Tank as a Raw Material Base for Obtaining Low Calorie Gaseous Fuel

Andrii Toporov and Olha Aleksieieva

Abstract The most important modern task is to reduce of the energy costs. It can be achieved in two ways—either by radical modernization of existing equipment, or by using alternative cheap energy resources. Radical modernization will require significant financial costs. The search and use of alternative cheap energy resources currently seems to be more promising. It is possible to use sludge from domestic settling tanks as a low-grade fuel alternative to gas and coal. The use of them has a number of advantages: low cost, fairly large quantity accumulated in the filtration fields of wastewater treatment plants, as well as those generated continuously; releasing land occupied for waste, reducing the level of environmental pollution; the possibility of using the existing infrastructure of processing enterprises for the construction of industrial installations for processing sludge from domestic settling tanks. Direct combustion of sludge from domestic settling tanks as a solid fuel is associated with significant difficulties, therefore it is advisable to convert the sludge as a solid fuel into gaseous fuel at first. However, in order to gasify sludges, they must have a set of physical and mechanical properties and their preliminary preparation is required. The preparation like that is briquetting. The work presents a study of the moisture content of the original sludge, an analysis of the drying dynamics, their structural, compression and strength characteristics, and established the features and patterns of behavior during briquetting. Recommendations are given for carrying out the processes of drying and briquetting sludge from domestic settling tanks.

Keywords Sludge from domestic settlement tanks · Briquetting · Drying · Structural · Compression · Strength characteristics of sludge

A. Toporov (✉) · O. Aleksieieva
Donetsk National Technical University, Pokrovsk, Ukraine
e-mail: andrii.toporov@donntu.edu.ua

O. Aleksieieva
e-mail: olha.aleksieieva@donntu.edu.ua

P. S. Brito et al. (eds.), *Insights into Energy Trends*, Green Energy and Technology,
https://doi.org/10.1007/978-3-031-83811-8_3

1 Introduction

Despite the fact that gasification technologies are quite well known [1], intensive research continues throughout the world. Thus, much attention of researchers is devoted to expanding the raw material base [2], special attention is paid to the gasification of various industrial [3] and household wastes [4]. Currently, when gasifying waste, researchers focus on obtaining hydrogen from synthesis gas [5–7]. While improving gasification technology, processes are optimized [8], in order not only to improve efficiency [9], but also to improve environmental performance [10, 11].

An assessment of the amount of sludge generated can be made based on the results of the work of already functioning wastewater treatment plants in 2 cities with a population of about 150,000 people. Approximately, the amount of main precipitation can be determined using the initial value of the release of dry suspended solids by one person per day, approximately this value is for 35–50 g, and the amount of silt formed in terms of dry mass is approximately 6 t per day. With a common sewage system with storm drains that allow to increase the amount of treated wastewater for two times, the amount of precipitation increases by 1.6–2 times.

The sludge ponds of one of the researched cities have a total size of 700 × 300 m^2, with a sediment thickness of ~4 m, i.e. the volume of wet sludge is ~840,000 m^3. With a humidity of ~70% and a density of dry sludge of ~1100–1200 kg/m^3, the mass of dry accumulated sludge is ~280,000 tons. With a consumption of ~200 tons/day, during the heating season this amount will be enough for ~7 years of operation. This is without taking into account new- formed sediment, the dry mass of which is ~15–20 tons/day on daily volume.

In the 2nd city under study there are 11 sludge ponds with dimensions of 60 × 70 m, with a sediment thickness of ~0.6 m, i.e. the volume of wet sludge is ~27,000 m^3. With a humidity of ~70% and a density of dry sludge of ~1100–1200 kg/m^3, the mass of dry accumulated sludge is ~10,000 tons. With a consumption of ~200 tons/day, during the heating season this amount will be enough for ~0.5 years of operation. This does not take into account newly formed sediment, the dry mass of which is ~1.5–2.0 tons/day.

2 Characteristics of Sludge from Domestic Settling Tanks

Dirty impurities retained by treatment facilities are found in wastewater in dissolved and undissolved form. The relative amounts of these and other impurities in wastewater are almost the same. The content of organic substances in undissolved impurities is almost twice higher than in dissolved ones.

It is practically possible to separate impurities into only three fractions—sediments, colloids, and dissolved impurities. It is explained by the fact that part of the suspension falls out along with the sediment, and the rest part is extracted along with colloids by adsorption or coagulation. For domestic wastewater, the ratio of the above three fractions can be expressed on average as 1:1:1, i.e. 33% for each fraction.

Undissolved impurities, depending on their different specific gravity, occupy different positions in a moving stream of water. Here, heavy impurities, mainly of mineral origin, move along the bottom of the channel. These impurities are usually retained by sand traps.

Floating impurities, as the lightest, occupy the top position. There are fats and fibrous substances. These impurities usually float to the surface in settling tanks or are retained by fat traps.

An intermediate position in the channel is occupied by suspended impurities, mainly of organic origin, called sediments. The volume of these impurities is 30–60 times greater than that of heavy or floating impurities. Suspended impurities, or sediments, are usually retained by primary settling tanks.

As for dissolved impurities contained in wastewater, the latter, as a result of biological treatment, are released in the form of a small flaky watery mass formed by bacteria and those smallest impurities that were not retained by mechanical treatment facilities.

These impurities differ in many respects in composition and structure from undissolved ones—therefore it is more convenient to call them "sludge", "active sludge". The last one is usually retained by secondary settling tanks after biofilters, aerofilters, and aeration tanks.

Sediment or sludge subject to fermentation (rotting) in various structures (septic tank, two-tier settling tank, methane tank) is called "fermented sludge" or "fermented sludge".

Thus, we can give the following classification of all mud impurities in wastewater, mainly according to their physical characteristics:

- hard impurities or waste retained by the gratings; heavy impurities, or impurities retained by sand traps;
- floating impurities, or impurities that float to the surface in settling tanks or are retained by grease traps;
- suspended impurities, or sediments (fresh sediments), retained by primary settling tanks;
- dissolved and colloidal impurities, or "sludge", "activated sludge", retained by secondary settling tanks after biofilters, aerofilters, aeration tanks;
- digested sludge, digested sludge, or a digested mixture of sludge from primary settling tanks and sludge from secondary settling tanks.

It should be noted that the composition and quantity of individual types of impurities listed in the classification are very diverse and depends mainly on the composition and quantity of wastewater, the sewerage system, the method of wastewater treatment, and the method of operation of treatment facilities.

The compound of coarse impurities or waste retained by the gratings is very variable. These are mainly large suspended and floating impurities, predominantly of organic origin. This includes kitchen waste (in the form of food residues), fibrous substances, paper, wood, etc. The amount of coarse sediment retained depends on the

Table 1 Sizes of sludge particles from domestic settling tanks

Particle size, mm	Number of particles, %	
	By weight	By volume
>8	0.04	0.32
5–8	0.05	0.44
3–5	0.17	1.03
2–3	0.22	1.51
1–2	1.37	14.00
<1	98.15	82.70

design of the grate and, in particular, on the size of the gaps between the rods, as well as on the composition of the waste liquid. The main part of dry sediment consists of organic matter. With a common sewage system, their number reaches 70%, and with a separate system—80% or more.

The moisture content of the sludge after settling wastewater for two hours is approximately 97%. During one or two days of staying in the settling tank, the sediments become compacted and their humidity at the bottom of the settling tank reaches 92–93%. The humidity of the sludge discharged from the settling tank increases to approximately 95%. It can be explained by the fact that, the part of the wastewater inevitably flows out along with the sludge, diluting the sludge when it is discharged. Sediments from primary settling tanks, having the properties of colloids, retain water well and, therefore, are difficult to dry. The sizes of sludge particles are as follows (Table 1).

The bacterial membrane that is formed in the body of biofilters and aerofilters is partially destroyed during the operation of these structures and enters with water into a secondary settling tank, where it is deposited in the form of sludge. This sludge consists of impurities consisting of aerobic bacteria and tiny particles that are not retained by the primary settling tanks. Unlike fresh sediments, this sludge is characterized by a higher stability of the organic substances contained in it and in this respect is close to activated sludge from aeration tanks. Its transition to full-fledged activated sludge requires aeration for 1–2 days, while fresh sediment would require aeration for 1–1.5 months. With its mechanical composition, sludge from secondary settling tanks is a thin suspension with particle sizes less than 1 mm. The sludge discharged from secondary settling tanks contains approximately 96% water.

Activated sludge from secondary settling tanks after aeration tanks is an amorphous, flak-like mass of brown color, richly populated by aerobic bacteria and other organisms. In fresh condition, sludge has almost no smell or has an earthy odor, but when it starts rotting, it emits a specific, unpleasant odor. Activated sludge differs with adsorbing and coagulating properties and has a great ability to mineralize organic substances contained in wastewater. The sorption surface of activated sludge flakes is quite large. In 1 ml of liquid with sludge taken from an aeration tank, there are about 100 million bacteria. The total surface of these bacteria in 1 m^3 is about 1200 m^2. This surface adsorbs colloids and suspensions from wastewater quite well.

In terms of mechanical composition, activated sludge refers to thin suspensions with particle sizes less than 1 mm. A characteristic feature of activated sludge is its high content of water. When pumping from secondary settling tanks, the moisture content of sludge reaches 99–99.5%. Activated sludge is easily agitated and quickly settles when settled. After 2–3 h of settling, it is compacted to 97–98% humidity.

In secondary settling tanks, along with recycled sludge, excess activated sludge falls out, resulting from wastewater treatment in aeration tanks. The amount of excess activated sludge that has to be dealt with when removing, processing and using sludge is quite significant and is taken to be 1 L/day per person or more.

Thus, sewage sludge is a dispersed system, where the dispersed phase is solid particles, and the dispersed medium is wastewater with electrolytes and organic substances dissolved in. Such system is called a suspension. The properties of this suspension largely depend on the moisture content of the sediment or, what is the same, on the water content in the sediment. Thus, the volume of precipitation, plastic-viscous properties and other physical and mechanical properties of precipitation depend on humidity.

The humidity of the bulk of precipitation usually ranges from 85 to 99%. For a more correct understanding of water in sediments, three types of water should be distinguished: free, colloidally bound and hygroscopic.

With a further decrease in humidity below 65–75%, water leaves the sediment, reducing its weight, but the volume, as experiments show, does not decrease due to the elasticity of the solid particles of the sediment.

3 Main Determined Properties of Municipal Waste Sludge

During the thermochemical processing of solid fuel, the technology of its processing is determined by the various properties of the raw material. Depending on humidity, ash content, particle size distribution, yield of thermochemical destruction products, ash composition, fundamentally different processing technology schemes and equipment are used for different types of raw materials, even with the same calorific value. Therefore, when developing a technology for producing flammable gas from sludge from domestic settling tanks, in addition to the main process of thermochemical processing, it is necessary to study the following sludge properties:

- determining the moisture content of the initial sludge;
- analysis of drying dynamics;
- research into briquetting of raw materials by conducting compression, shear and strength tests.

Table 2 Dynamics of sample drying

Drying time, hour	Weight loss according to samples, %						
	1	2	3	4	5	6	7
0	0.00	0.00	0.00	0.00	0.00	0.00	0.00
2	35.10	40.24	40.24	30.19	30.19	29.19	25.22
3	42.05	51.25	50.00	36.91	49.70	36.32	32.50
4	42.10	60.11	55.11	37.01	50.15	39.11	33.02
5	42.10	67.01	58.07	37.01	50.33	39.70	33.30
6	42.10	70.24	58.10	37.01	50.33	39.70	33.30
7	42.10	74.20	58.10	37.01	50.33	39.70	33.30
8	42.10	74.20	58.10	37.01	50.33	39.70	33.30

3.1 Determination of the Moisture Content of the Original Sludge

Moisture content was determined for all sludge samples. Due to the fact that the source material, sludge, had an extremely heterogeneous structure, high humidity, and a tendency to stick together and clump, it was impossible to grind it to the required analytical sample size of <0.2 mm using conventional laboratory methods. Therefore, the determination of analytical moisture was examined as follows. A sample of sludge weighted ~15 g was placed in an aluminum bottle and evenly distributed throughout the volume of the bottle with a layer thickness of 25 mm. Then the bottle with the sludge sample was placed in an electrically heated drying cabinet operating at atmospheric pressure. The temperature in the drying cabinet was maintained at 105 ± 5 °C. After 2 h, the bottle was removed, cooled and weighted on a scale with an accuracy of 0.01 g, then again placed in a heated cabinet. All the next weightings were carried out with the intervals of 0.5 h until the difference between the last two weightings was no more than 0.1%. The lowest mass was taken as the result. The results of drying dynamics are presented in Table 2.

3.2 Determination of the Moisture Content of the Original Sludge

The granulometric composition of the sludge was determined by sieving air-dried samples on sieves with round holes. The studied sludge samples are represented mainly by particles with a size of less than 1 mm is 53.9–61.2%. The results of determining the granulometric composition are given in Table 3.

Table 3 Granulometric composition of the original sludge

Sample number	Number of particles, %, size, mm				
	>8	5–8	5–3	1–3	<1
1	10.4	7.2	8.3	20.2	53.9
2	5.2	3.8	4.1	25.7	61.2
3	10.1	6.0	7.2	17.2	59.5
4	7.9	4.1	5.5	22.3	60.2
5	5.0	4.4	5.1	27.5	58.0
6	5.1	5.3	7.5	24.8	57.3
7	5.9	4.9	8.1	23.9	57.2

3.3 *Determination of Compression, Shear and Strength Properties of Silts*

Briquetting sludge will increase its bulk mass, safe transportation costs, and also reduce dust formation when processing it in furnaces with internal heating by a gaseous coolant. To determine the parameters of the briquetting process, it is necessary to conduct compression, shear and strength tests, and solve the following problems:

- to determine the fundamental possibility of sludge compaction;
- to determine the optimal pressing pressure of briquettes, sufficient for subsequent processing of their strength, with minimal pressing energy costs;
- to determine the specific energy consumption for pressing.

Compression properties characterize the ability of a bulk material to change density depending on the pressure applied to it. The principle of compression tests is that when a material—sludge is compressed—its deformation is controlled, which makes it possible to calculate the parameters of the compression curve.

To determine each parameter, a series of experiments consisting of 3–5 tests were carried out, for each of them were obtained the corresponding dependencies.

The processes occurring in the material while briquetting: mutual ordering, shape change and destruction of particles, displacement of gases and the liquid phase, significantly depend on the type of material under research. With a relatively rapid increase in pressure on the material, the relaxation time was 20–40 s, which is explained by the low rates of strain transfer, thus the load dropped to almost zero.

The compression properties of the pressed material (sludge) are mainly characterized by the coefficient K_y of its elastic expansion and the parameters of the compression curve.

Elastic expansion coefficient:

$$K_y = \rho(\sigma_{kappa})/\rho(\sigma_p) \quad (1)$$

where $\rho(\sigma_\kappa)$ is density of material at final compaction pressure, $\rho(\sigma_p)$ is density of the material after elastic expansion.

The compression curve, which characterizes the ability of a material to change density (volume) when the pressure applied to it changes, is described by the equation:

$$\rho = \rho_1 \sigma^n \tag{2}$$

where ρ_1 is a coefficient numerically equal to the density of the bulk material at unit pressure σ is average stress value at the point under consideration, n is an indicator of the degree of compression curve.

The strength properties of silts are mainly characterized by the compressive strength σ_{pr} compressive strength, tensile strength σ_{tn}, and the strength profile:

$$\tau = \tau_0 + \mathrm{tg}\,\varphi \cdot \sigma \tag{3}$$

where τ_0 is the cohesiveness of the bulk material, φ—angle of internal friction.

The compressed briquettes were tested for static compressive and tensile strength.

4 Results of Compression, Shear and Strength Properties of Municipal Waste Sludge

When studying sludge briquetting, the compression, shear and strength properties of household waste sludge were determined, which are shown in summary Table 4. As a result a series of tests, a number of parameters were determined (in brackets - the column number the table) with the parameter: compaction pressure (2), modulus elasticity (3), briquette density (4), coefficient of elastic expansion (5), specific energy intensity of pressing (6), density at unit pressure (7), exponent of the compression curve (8), briquette compressive strength (9), strength tensile briquette (10), material cohesion (11), slope angle of the Mohr envelope curve (12), tangent of the Mohr envelope curve slope angle (13).

When pressing briquettes, structuring of the material was observed, which consisted in the orientation of lamellar sludge particles perpendicular to the direction of action of the active pressing pressure (principal stresses).

After removing the load, local swelling was observed on the surface of the briquettes, associated with the presence of silt particles in the surface layers with significant elastic deformation. The height of elastic bulging of particles above the surface did not exceed 30% of the maximum size of sludge particles. Also, in addition to swelling, the formation of shallow annular cracks with a depth of 3–4 mm, passing through the middle part of the briquette, was also observed. The interface, as a rule, passed along sludge particles, the surface of which was poorly bonded to the sludge particles. Sludge adhesion on the surface of the rods and matrix was often observed.

Table 4 Summary table of the results of compression and strength tests

№	P, MPa	E, MPa	ρ, kg/m^3	K_y –	A, kJ/kg	Compression curve		Strength		Strength certificate		
						ρ_1, kg/m^3	n –	σ_{pr}, MPa	σ_{tn}, MPa	τ_0, kPa	φ, radian	tg(φ) –
1	2	3	4	5	6	7	8	9	10	11	12	13
1.	5.19	45.8	1199.1	1.08	0.436	1096.7	0.11	1.75	0.07	0.17	67.43	2.41
2.	10.35	93.0	1303.3	1.11	1.061	1090.5	0.11	2.60	0.15	0.32	62.63	1.93
3.	15.53	131.1	1310.2	1.15	2.091	1039.4	0.12	3.69	0.25	0.48	60.80	1.79
4.	5.19	46.7	1062.1	1.04	1.478	834.8	0.12	1.78	0.27	0.35	47.31	1.08
5.	10.35	38.8	1101.2	1.19	2.014	515.2	0.28	3.45	0.16	0.33	63.54	1.91
6.	5.18	61.0	1126.8	1.04	0.351	1034.7	0.09	0.99	0.11	0.16	53.85	1.37
7.	10.35	104.6	1148.0	1.09	1.454	949.0	0.10	2.05	0.23	0.35	52.72	1.31
8.	15.53	132.0	1168.2	1.15	3.353	883.6	0.12	3.05	0.25	0.44	58.02	1.60
9.	9.70	73.0	1585.0	1.10	0.970	1275.8	0.13	3.22	0.15	0.34	65.98	2.24
10.	15.53	131.9	1515.1	1.13	1.950	1173.3	0.12	3.55	0.38	0.58	59.60	1.88

The study of the energy intensity of sludge pressing is associated with determining the most efficient conditions for the process in terms of energy consumption. The research range for the density of the compressed load is from 1062.1 kg/m^3 to 1585.0 kg/m^3. The value of specific energy consumption was calculated as the specific work spent on obtaining the current density of sludge. Thus, the dependence of the increase in specific energy consumption in the process of compacting briquettes has been established.

Analysis of energy consumption for pressing at pressures of 5, 10 and 15 MPa showed a wide range of values: 0.35–1.48, 1.10–2.13 and 1.95–3.35 kJ/kg, respectively. The reason for the fluctuation in energy consumption for pressing lies in the different conditions of sludge formation (composition, storage time in sludge fields, humidity, presence of impurities, etc.). In accordance with this, the strength of samples from different settling tanks varies over the entire pressure range.

When repeated loading was applied, the density increased more rapidly at low energy inputs up to the density obtained at 10 MPa. The density of the first pressing was achieved with significantly lower loads. The relaxation time increased with each subsequent application of pressure, and the load dropped to only 1–3 MPa. This indicates that with repeated compaction, the material's properties approach a solid elastic body, for which the strain transfer rates are high.

With several pressings of the same material, a number of effects are observed: an increase in the final density and strength of the briquettes, a decrease in elastic expansion coefficients, lateral pressure coefficients and compaction energy intensity, which decreases by 50–60%.

The strength characteristics of briquetted mixtures depend on the composition of the mixtures, pressing pressure, humidity and other factors. For sludges after pressing to a pressure of 5 MPa, $\sigma_{pr} = 0.99$–1.75 MPa, $\sigma_{tn} = 0.07$–0.27 MPa were obtained, which allows us to consider briquettes as a brittle material. When the compaction pressure increases to 15 MPa, a significant strengthening of the material occurs ($\sigma_{pr} = 3.05$–3.69 MPa), while 2.8–6.4 more energy is spent on pressing (depending on the sample).

5 Conclusions

As a result of comprehensive experimental studies of the properties of sludge from domestic settling tanks, their structural, compression and strength characteristics were determined, and the following features and patterns of behavior during briquetting were established:

1. Silts have a number of advantages compared to other types of raw materials (peat, coal, oil shale, etc.)—the possibility of open-pit mining, the absence of stripping and drilling and blasting, the presence of the necessary transport infrastructure and the close location of possible processing plants.
2. From the nature of changes in the density and strength of briquettes it follows that the maximum pressing pressure should be 15 MPa, because its further increase leads to a slight increase in density and strength, but a sharp increase in energy consumption.
3. When compacting different sludge samples, the specific energy consumption differs slightly, which makes it possible to use a press with the same design parameters for compacting all sludges.
4. Upon re-compaction, a noticeable change in compression properties occurs. The density of briquettes increases by an average of 7%. The best results were obtained with lower binder content (14% increase in density). Moreover, the increase in strength is directly proportional to energy consumption.
5. The strength of briquettes obtained at a pressure of 15 MPa and natural drying for ~1 day is sufficient for transportation and 1–2 overloads in technological equipment.
6. The strength of briquettes after drying increases significantly, so it can be recommended to keep the briquettes (natural drying) for 2–5 days before transportation.

References

1. Peng D, Xiang X, Deng Z, Zhou X, Wang B, He C (2024) Study on emission factor and reduction potential of organic solid waste gasification process. Case Stud Thermal Eng 53:103978. ISSN 2214-157X. https://doi.org/10.1016/j.csite.2024.103978
2. Biancini G, Cioccolanti L, Moradi R, Moglie M (2024) Comparative study of steam, organic Rankine cycle and supercritical CO_2 power plants integrated with residual municipal solid waste

gasification for district heating and cooling. Appl Thermal Eng 241:122437. ISSN 1359-4311. https://doi.org/10.1016/j.applthermaleng.2024.122437

3. Saravanakumar A, Rajendran Sudha M, Pradeshwaran V, Lih Jie Ling J, Hoon Lee S (2024) Green circular economy of co-gasification with municipal solid waste and wood waste in a novel downdraft gasifier with rotating grate. Chem Eng J 479:147987. ISSN 1385-8947. https://doi.org/10.1016/j.cej.2023.147987
4. Amin N, Khan Z, Razzaq A, Ghauri M, Khurram S, Inayat A, Jaffery M, Hameed Z (2024) Municipal solid waste air gasification using waste marble powder as a catalyst for syngas production. J Energy Inst 113:101496. ISSN 1743-9671. https://doi.org/10.1016/j.joei.2023.101496
5. Makwana J, Dhass AD, Ramana PV, Sapariya D, Patel D (2027) An analysis of waste/biomass gasification producing hydrogen-rich syngas: a review. Int J Thermofluids 20:100492. ISSN 2666-2027. https://doi.org/10.1016/j.ijft.2023.100492
6. Darabadi Zare AA, Yari M (2024) Comprehensive examination and analysis of thermodynamics in a multi-generation hydrogen, heat, and power system based on plastic waste gasification integrated biogas-fueled chemical looping combustion. Process Saf Environ Protect 183:976–991. ISSN 0957-5820. https://doi.org/10.1016/j.psep.2024.01.060
7. Zhang P, Liang C, Wu M, Li Y, Chen X, Liu D, Ma J (2024) Sustainable microwave-driven CO2 gasification of plastic waste for high-yield H2 and CO production. Appl Catal B: Environ Energy 345:123718. ISSN 0926-3373. https://doi.org/10.1016/j.apcatb.2024.123718
8. Tomasek S, Bárkányi Á, Egedy A, Miskolczi N (2024) Model-based determination of optimal operating parameters for different solid waste gasification. Chem Eng J Adv 17:100586. ISSN 2666-8211. https://doi.org/10.1016/j.ceja.2024.100586
9. Guo Y, Hu W, Feng G, Zhao Y, Li C, Wang X, Ma J (2024) Study on the excitation effect and mechanism of coal gasification slag based on solid waste. Powder Technol 435:119460. ISSN 0032-5910. https://doi.org/10.1016/j.powtec.2024.119460
10. Xiao Z, Wu W (2024) Environmental sustainability by an integrated system based on municipal solid waste fluidized bed gasification using energy, exergy, and environmental analyses. Appl Thermal Eng 239:122171. ISSN 1359-4311. https://doi.org/10.1016/j.applthermaleng.2023.122171
11. Sieradzka M, Mlonka-Mędrala A, Błoniarz A, Magdziarz A (2024) Experimental study of biomass waste gasification: impact of atmosphere and catalysts presence on quality of syngas production. Bioresour Technol 394:130290. ISSN 0960-8524. https://doi.org/10.1016/j.biortech.2023.130290

Software-Based Fluid Dynamic Simulation of a Combustion Chamber

Alejandro Martínez-Martín, Alfonso Carlos Marcos Romero, Pedro José Muñoz Acebedo, and Juan Félix González González

Abstract Computational fluid dynamics (CFD) offers a series of advantages, mainly in terms of speed and capacity to obtain results that are faithful to the real model. This work details the process to be followed to carry out a CFD simulation of the combustion process in a combustor. The guidelines to be followed from the initial stage of generating the initial geometry to obtaining the results of the combustion produced inside the chamber are established. The critical aspects that must be considered during the process are highlighted, as well as the elements that posed the main difficulty for the authors during the simulation. This can be used as a guide by other authors who wish to carry out a simulation of a combustion chamber or simulations of similar characteristics.

Keywords Simulation · CFD · Combustion · Burner · Flame holder · Combustion chamber · Autodesk inventor · Ansys fluent

1 Introduction

Computational fluid dynamics (CFD) offers multiple advantages in terms of speed, accuracy and reliability. It is currently widely used in different research fields, allowing a significant reduction in time and resources with respect to laboratory

A. Martínez-Martín (✉) · P. J. Muñoz Acebedo · J. F. González González
Department of Applied Physics, Industrial Engineering School, University of Extremadura, Badajoz, Spain
e-mail: alejandromm@unex.es

P. J. Muñoz Acebedo
e-mail: pjmunoza01@unex.es

J. F. González González
e-mail: jfelixgg@unex.es

A. C. Marcos Romero
Department of Graphic Expression, Industrial Engineering School, University of Extremadura, Badajoz, Spain
e-mail: acmarcos@unex.es

P. S. Brito et al. (eds.), *Insights into Energy Trends*, Green Energy and Technology,
https://doi.org/10.1007/978-3-031-83811-8_4

experimentation. The results obtained in this type of simulations show a high level of accuracy and similarity with respect to reality. This work is the continuation of previous work carried out by the research group [1, 2].

The followed steps for the simulation development are listed below:

(1) Initial model definition. Geometry parameterisation.
(2) Control volume generation. Exportation to calculation program. Meshing.
(3) Combustion modelling.
(4) Boundary conditions setting.
(5) Resolution process configuration.
(6) Results analysis.

2 Model Definition

In this section, the steps taken for the parameterisation of the simulated model are detailed.

2.1 Geometry Parameterisation

First, an initial model of the combustor was defined. This model was used as a starting point to be subsequently modified according to the results obtained from the successive simulations. In order to define the geometric dimensions, the formulas and procedures provided by Lefebvre [2], a renowned author in the field, and widely used in several previous studies with similar characteristics [3–6], were used. Once the dimensions of the chamber had been determined, a three-dimensional model was created using computer-aided design (CAD) software, in this case Autodesk Inventor (Fig. 1). The main advantage of using this type of software is the ability to parameterise the model. This means that the key variables for the design (chamber length, diameter, distribution and dimension of the burner orifices) can be set as independent input variables and then easily modified according to the results obtained in the simulations, without having to rebuild the whole model from scratch.

At this point it is essential to generate a closed geometry, otherwise computational problems will arise during the calculation phase. If the camera is symmetrical about its axial axis (Fig. 2a), it is possible to generate a two-dimensional model corresponding to a planar section of the camera and perform the calculations on it (Fig. 2b), as the results will be symmetrical about the camera axis. The main advantage of this approach is a significant reduction in computation time and easier meshing.

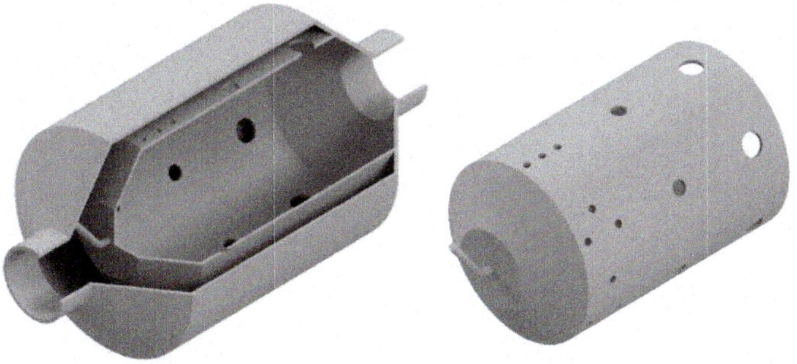

Fig. 1 Isometric view of the chamber (left) and the inner flame holder (right)

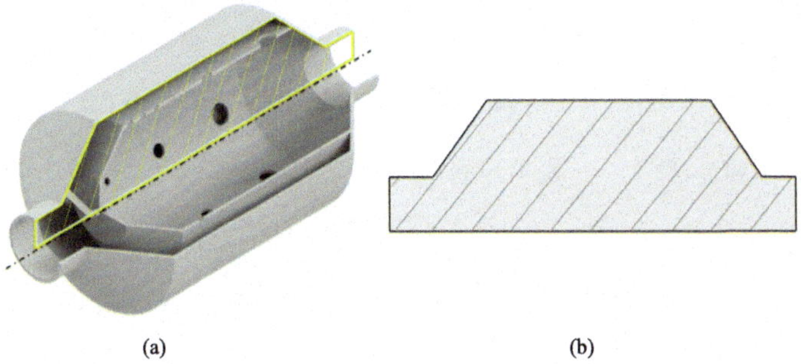

Fig. 2 Symmetry of the chamber about its axial axis (**a**) and two-dimensional chamber section (**b**)

2.2 Control Volume Generation

After establishing the three-dimensional model of the chamber, it was necessary to generate a control volume of the fluid to be simulated inside the chamber. This represents the volume occupied by the fluid. Calculation programmes often have the option to generate this volume directly. Alternatively, it can be defined manually by subtracting the volume of the chamber from a solid, similar to the process of creating a mould. This is illustrated in Fig. 3.

To simulate fluid dynamics, the chamber's control volume was exported to computational fluid dynamics (CFD) software in STL format, which is widely used in rapid prototyping. The simulation was performed using Ansys Fluent. The imported geometry was used to generate a three-dimensional mesh, which involved carrying out a finite element discretisation of the model. The quality of this mesh is critical to the accuracy and convergence of the simulation results. Depending on the complexity of the model to be simulated, the mesh can be either structured (*i.e.* the generated mesh will follow a regular distribution in rows and columns, Fig. 4a) or unstructured (on the contrary, the mesh will follow an irregular pattern, Fig. 4b, c). Structured meshing offers the main advantage of requiring fewer computational resources during the computational phase. This considerably reduces resolution times, resulting in more

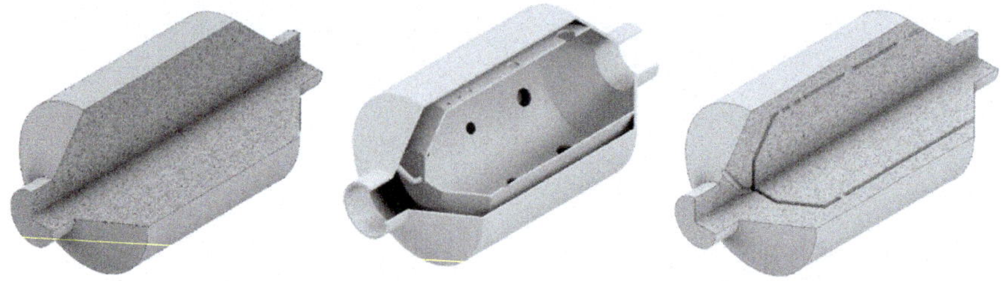

(a) (b) (c)

Fig. 3 Elements used to generate the volume within the chamber: Solid body (**a**), combustion chamber geometry (**b**) and control volume (**c**)

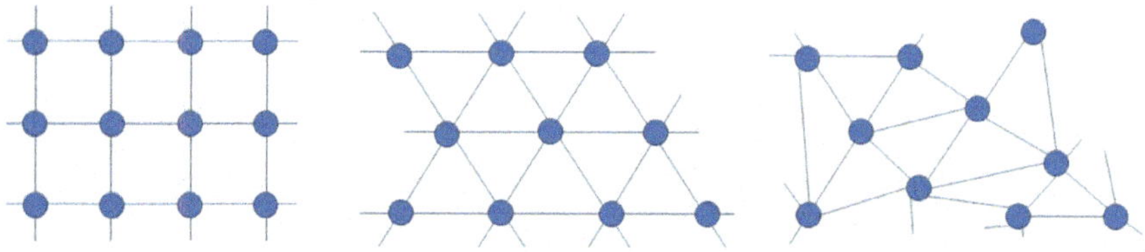

(a) (b) (c)

Fig. 4 Structured (**a**) and unstructured (**b** and **c**) mesh geometry

efficient and accurate solutions. However, certain complex geometries may make it impossible to generate this type of mesh.

The Ansys Meshing program has an automatic mesh generation option, although a subsequent refinement stage is necessary, *i.e.*, the establishment of a greater number of nodes in those areas that are critical for the simulation. In this work, when simulating a combustion chamber, the mesh density was increased in those areas where there was a significant temperature gradient, for example, in the areas close to the fuel injector where the flame was generated and on the walls of the chamber.

To facilitate the subsequent establishment of the boundary conditions, it was necessary to define those areas where fluid enters or leaves the control volume. Specifically, the section where air enters the chamber (inlet), fuel enters (fuel_inlet) and flue gas exits (outlet) was named. These sections are shown in Fig. 5.

3 Simulation Configuration

The following section details the configuration used in the simulation models.

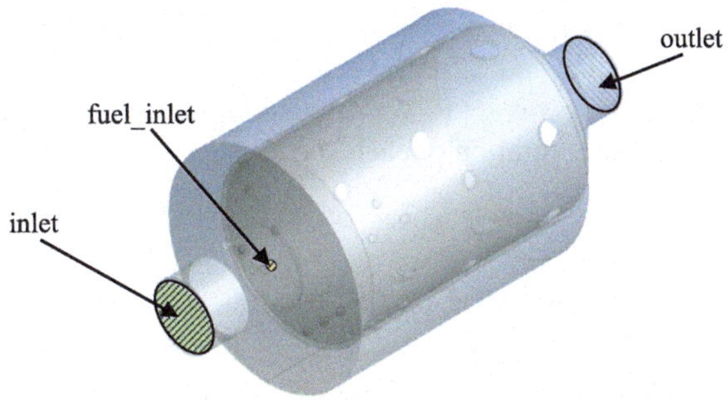

Fig. 5 Control volume sections

3.1 Combustion Models

The following section details the configuration used in the simulation models. Firstly, the energy equation was activated to analyse the heat exchange of the system, which enables the calculation of the temperature distribution inside the chamber. Secondly, the viscous model was activated due to the fluid flow inside the chamber. The viscosity of the fluid affects the velocity distribution, which allows for a more accurate representation of its behaviour. The simulation employed the k-epsilon turbulence model, which uses two transport equations to predict the turbulence variables k (specific kinetic energy) and epsilon (specific energy dissipation). This model is less computationally demanding and is recommended when the meshing has not been refined, striking a balance between result accuracy and computational time.

The species transport model was used to simulate the transport and reaction of the chemical species present. In this work, the combustion of methane inside the chamber was simulated by selecting the methane-air mixture. The option for volumetric reaction was enabled, which allows the chemical reactions to be modelled directly in the control volume, instead of basing the reaction on the convective and diffusive transport of the species present, thus obtaining a more detailed simulation. It is important to consider that the turbulence of the fluid can have a significant impact on the chemical reactions that occur during ignition and flame propagation within the combustion chamber. Therefore, the Eddy-Disipation Turbulence-Chemistry Interaction (EDTCI) model was used to account for this effect.

3.2 Materials

Fuel properties and chemical reaction rates were established in the materials section. The methane combustion reaction carried out is shown in Eq. (1).

$$CH_4 + 2O_2 \rightarrow CO_2 + 2H_2O \quad (1)$$

Within the species, it was necessary to define the stoichiometric coefficients (1 for methane and 2 for oxygen, as shown in Eq. 1). As for the exponent of the rate equation, it was set to 1 for both compounds, since the reaction rate is proportional to the concentration of each species.

3.3 Boundary Conditions

The inlet section established the air inlet conditions, including the air temperature set at room temperature, the fluid inlet speed, and the turbulence intensity set at 10%. The species present in the air were also defined, with a specific value of 0.23 for oxygen (as air contains 23% oxygen). The program considers the remaining percentage as nitrogen by default, so there was no need to define it. Similarly, the fuel inlet conditions (fuel_inlet) were established, with a value of 1 set for methane, indicating a pure methane inlet stream. The chamber walls were assumed to be at room temperature at the start of the simulation. During simulation development, the fuel and air inlet velocities were adjusted to analyse their impact on the reaction's behaviour.

4 Resolution Configuration and Results

Prior to solving the flow equations, defining the initial conditions for flow variables, including velocity, pressure, temperature, and fractions of different chemical species, is necessary. This establishes a starting point for successive iterations, aiding result convergence. A standard initialization was chosen in this case. However, selecting a hybrid initialization can be beneficial if real experimentally measured starting data or data obtained from previous simulations are available, as it may help the solution to converge more quickly.

In terms of the resolution method, a second-order dissipation ratio was chosen to achieve higher simulation accuracy. The following results were obtained: temperature profile (Fig. 6), pressures (Fig. 7), fuel concentration (Fig. 8), and particle trajectory (Fig. 9). Inlet velocities of 100 m/s and 20 m/s were used for the outside air (coming from a turbocharger) and the fuel, respectively. The velocities were chosen based on previous results and may be adjusted in future simulations depending on the specific analysis conditions.

The combustion of the mixture begins near the fuel injection point, causing a simultaneous temperature increase (as shown in Fig. 6) and a decrease in the concentration of the fuel mass fraction (as shown in Fig. 8). The flame reaches maximum temperatures of 1317 °C (1590 K), while the outer walls of the chamber reach 787 °C (1070 K). The pressure (as shown in Fig. 7) reaches a maximum of 1.01 bar, which must be supported by the chamber walls. Figure 9 displays the trajectory of the

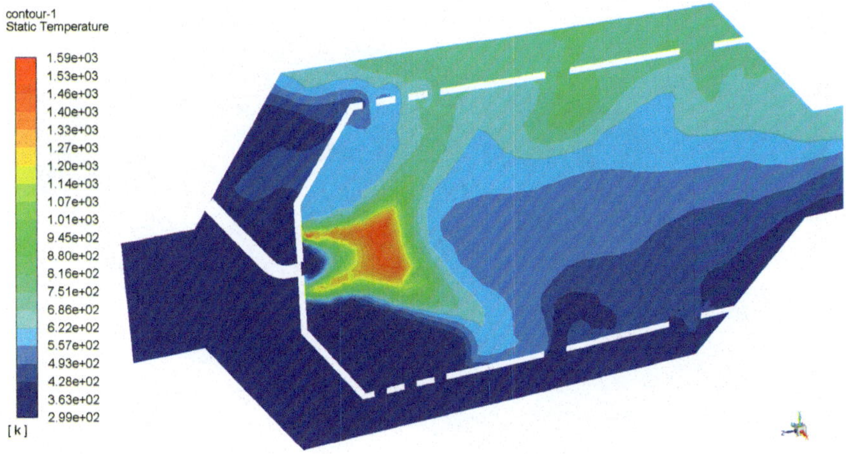

Fig. 6 Temperature profile in the chamber cross section

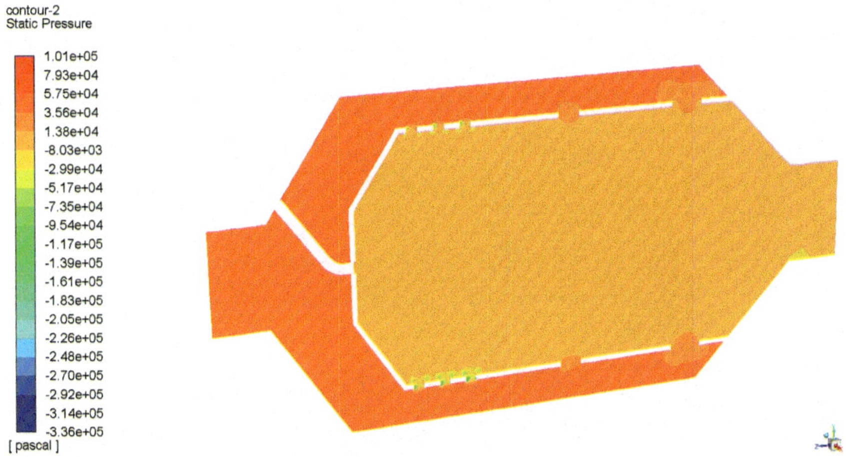

Fig. 7 Pressure profile in the chamber cross section

incoming air particles throughout the simulation. The dark blue colour indicates the start time of the simulation, which corresponds to the inlet air (inlet).

5 Conclusions

In this paper the fundamental aspects to be considered during the simulation phase of a combustion chamber have been pointed out. This work can be used as a guide by other authors who wish to carry out simulations with similar characteristics. Firstly,

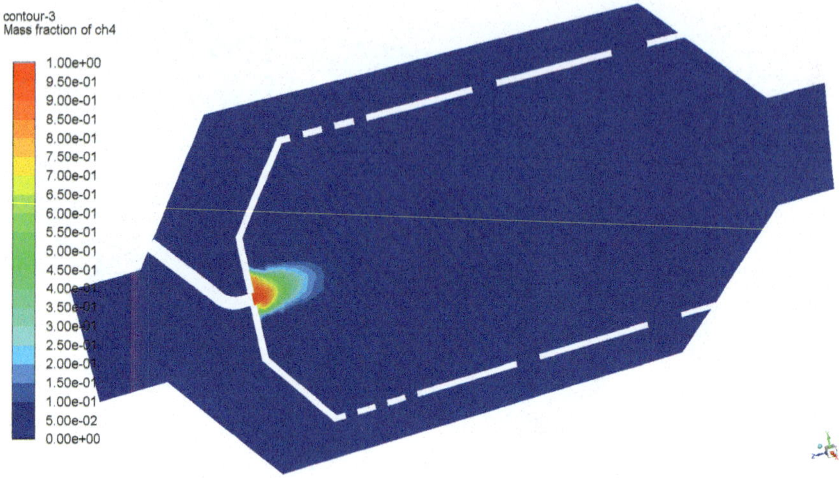

Fig. 8 Fuel concentration (CH_4) in the chamber cross section

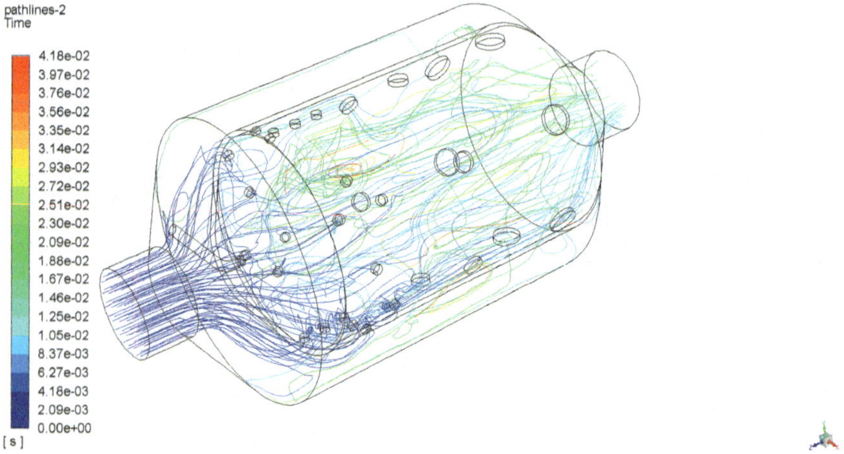

Fig. 9 Incoming air particle trajectory

the process of creating the three-dimensional model of the chamber was detailed, as well as the generation and export of the corresponding control volume, which requires a process of meshing and refinement. The key points for setting the conditions and parameters of the simulation have been provided, including: modelling the combustion, establishing the boundary conditions, defining the material properties and configuring the Solver to achieve convergence of the results. Using the Ansys Fluent programme, a complete simulation was carried out with the previously defined conditions, obtaining different profiles of temperatures, pressures, fuel concentration and particle trajectory, among others.

References

1. González-González JF, Alkassir A, San José J, González J, Gómez-Landero A (2014) Study of combustion process of biodiesel/gasoil mixture in a domestic heating boiler of 26.7 kW. Biomass Bioenergy 60:178–188
2. Gutiérrez J, Galán CA, Suárez R, Álvarez-Murillo A, González-González JF (2018) Biofuels from cardoon pyrolysis: extraction and application of biokerosene/kerosene mixtures in a self-manufactured jet engine. Energy Convers Manag 157:246–256
3. Lefebvre AH, Ballal Dilip R (2010) Gas turbine combustion, alternative fuels and emissions, 3rd edn.
4. Vilag VA, Vilag JA, Carlanescu R, Mangra AC, Florean FG,(2019) CFD application for gas turbine combustion simulations. In: Computational fluid dynamics simulations
5. Kulshreshtha DB, Channiwala SA (2005) Simplified design of combustion chamber for small gas turbine applications. In: Proceedings of the ASME 2005 international mechanical engineering congress and exposition. energy conversion and resources
6. Silva R, Lacava P (2013) Preliminary design of a combustion chamber for microturbine based in an automotive turbocharger. In: 22nd international congress of mechanical engineering

Study of Corn Crop Residue Pelletising in a Semi-industrial System

R. García-Mateos, M. T. Miranda, J. I. Arranz, P. Romero, F. J. Sepúlveda, and S. Cuellar-Borrego

Abstract Within the framework of a green and circular economy, the valorisation of agricultural residues plays a fundamental role, particularly in highly extended crops such as corn. In particular, such valorisation becomes especially important when the residue has a subsequent use as biofuel. In the present work, a feasibility study has been carried out on the production of pellets from corn crop residues for consumption in domestic boilers, using a semi-industrial pelletizing system. For this purpose, and starting from the specific residue of the corn cob waste, different scenarios of mass flow input of the residue and water input during the pelletizing process have been tested. After the tests carried out, it was concluded that it is necessary to add water during the process in order to obtain type A pellets according to the ISO quality standard. It has also been detected that there is an optimum mass flow to produce high quality pellets, and that values below or above the established range provide insufficient quality.

Keywords Valorisation · Corn residue · Pelletisation

1 Introduction

Currently, corn is one of the main crops worldwide, serving as a fundamental source of human nutrition alongside wheat and rice [1]. To grasp the magnitude of the importance of this crop, and based on data from the past 10 years provided by the Food and Agriculture Organization (FAO), it is observed that corn cultivation has progressively increased both in terms of production and cultivated hectares, reaching 1.163 billion tons produced and 203 million hectares cultivated in 2022 [1].

R. García-Mateos (✉) · S. Cuellar-Borrego
Scientific and Technological Research Centre of Extremadura (CICYTEX), Mérida, Spain
e-mail: raquel.garcia@juntaex.es

M. T. Miranda · J. I. Arranz · P. Romero · F. J. Sepúlveda
Department of Mechanical, Energy and Materials Engineering, Industrial Engineering School, University of Extremadura, Badajoz, Spain

P. S. Brito et al. (eds.), *Insights into Energy Trends*, Green Energy and Technology,
https://doi.org/10.1007/978-3-031-83811-8_5

Given the significance of this crop, within the framework of pursuing a green and circular economy, the valorisation of agricultural residue, specifically those from corn cultivation, becomes crucial. This is not only economically pivotal as a differential point for corn producers but also environmentally important as it transforms a residue into a usable resource. At the European level, the utilization of agricultural residues for use as bioenergy is included in the waste hierarchy established in Directive (UE) 2023/2413 of the European Parliament and of the Council of 18 October 2023, as a measure to ensure that energy from biomass is produced in a way that minimises undue distortive effects on the biomass raw material market and an adverse impact on biodiversity, the environment and the climate. In the case of agricultural residues, reuse and recycling prevail over bioenergy according to the hierarchy established in the Directive, but both options are not currently feasible for corn residues. In the case of corn crop residues, since it is a conversion of a residue into biomass, it is not affected by indirect land use change (ILUC), since the main objective of the crop is its food use, not its use as biofuel, and therefore there is no increase in carbon emissions.

Among the options for the valorisation of agricultural residue, the conversion of these residues into solid biofuels holds significant importance, with pelletisation being economically advantageous for their potential use in domestic boilers.

During corn harvesting, a residue known as stover is produced, consisting of the stalks and leaves of the plant, along with the shelled cobs. The interest in valorising this residue has been previously explored, both collectively and separately, for the corncob waste and stalk. Miranda et al. [2] studied the ash content of corn residues, obtaining lower percentages in those residues that were mainly composed of corn cob waste. This parameter is considered fundamental in the quality of pellets for household consumption, so in the present study the pelleting of corn cob residues is carried out.

The objective of this study is conduct pelletisation trials of corncob waste from corn cultivation using a semi-industrial pelletisation system. This system allows modifying the input conditions to determine the most optimal operational scenarios for producing pellets for domestic consumption. The quality of the pellets produced will be tested according to the applicable ISO standard for non-wood pellets.

2 Materials and Methods

To conduct the experiments, the chosen raw material consisted of corn cob waste along with its surrounding leaves. These materials were utilized to investigate the pelletisation process subsequent to the selective collection of the residue.

The samples underwent grinding procedures utilizing a hammer mill (CIP Line SG40, Italy) equipped with a 5 mm sieve at the machine's discharge outlet, in order to achieve the desired particle size for further pelletisation.

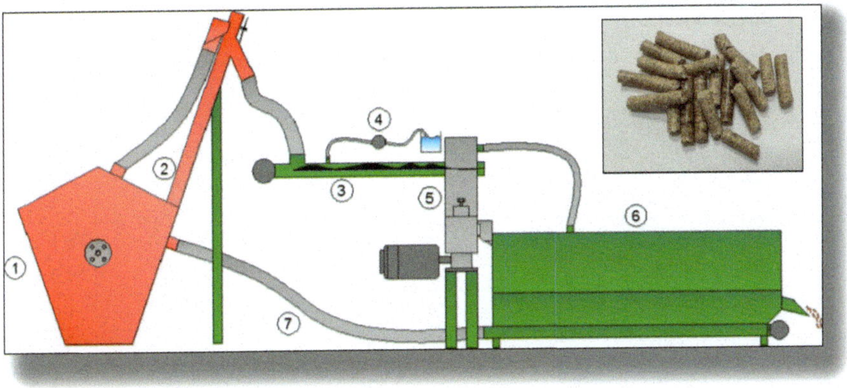

Fig. 1 Scheme of the pelletiser [9]

The crushed sample was characterised prior to pelleting by proximate and ultimate analysis and determination of the higher heating value. For this purpose, the established ISO standards were used [3–8].

Densification tests were performed in a pelletizing machine Kovo Novak MGL 200, provided with rotary flat array with 6 mm diameter holes. The rollers rotate about their longitudinal axis on the surface of the flat die, thereby exerting elevated pressure on the product and compelling it through perforations, facilitating compaction.

Figure 1 shows a schematic diagram of the pelletiser [9], where: the raw material is placed in the hopper (1), being carried by an auger feeder (2 and 3). During this process, the raw material is mixed with water through the moisture-control system (4). The resulting mixture passes to the flat die (5), where it is extruded. Finally, the resulting pellets are cooled down in the cooling drum (6), and the surplus product returns to the hopper (7).

The test was carried out on the basis of a raw material with an average moisture content of 12.76%. This initial moisture content is considered optimal for pelletising processes according to the literature [10, 11].

Four different tests were carried out, varying the mass flow rates of raw material fed to the pelletiser inlet. In addition, and for a given mass flow of raw material, seven different scenarios of moisture input to the raw material were carried out. The test scenarios chosen were based on pelleting experience, as low mass flows do not provide sufficient material for the compaction process to take place, while excessive mass flows cause the equipment to collapse. As for the water input, low water contents do not allow pelletising by greatly increasing the temperature of the flat die of the pelletiser, while with excessive inputs the necessary compacting of the pellet does not take place and the pellets are broken down.

Table 1 summarises the analyses performed, including the ratio of water flow provided to mass flow, represented by the parameter R.

The pellets obtained were subjected to various characterization tests. During pelleting, samples were collected at intermediate stages of production for 2 min, excluding pellets produced in the initial and final stages. For each of the pelleting assays, the average moisture content of the pellets produce was determined according

Table 1 Test conducted

Test number	Mass flow rate (g/min)	Water flow provided (g/min)	R
A1	302	3.19	10.6
A2	302	3.76	12.5
A3	302	4.40	14.6
A4	302	5.23	17.3
A5	302	5.48	18.1
A6	302	5.72	18.9
A7	302	6.53	21.6
B1	462	3.19	6.9
B2	462	3.76	8.1
B3	462	4.40	9.5
B4	462	5.23	11.3
B5	462	5.48	11.9
B6	462	5.72	12.4
B7	462	6.53	14.1
C1	666	3.19	4.8
C2	666	3.76	5.6
C3	666	4.40	6.6
C4	666	5.23	7.9
C5	666	5.48	8.2
C6	666	5.72	8.6
C7	666	6.53	9.8
D1	902	3.19	3.5
D2	902	3.76	4.2
D3	902	4.40	4.9
D4	902	5.23	5.8
D5	902	5.48	6.1
D6	902	5.72	6.3
D7	902	6.53	7.2

to the standard ISO 18134-2:2017 [5], as well as the durability test according to the standard ISO 17831-1:2015 [12]. The average sizes (diameter and length) were determined using a calliper according to ISO 17829: 2015 [13].

Subsequently, the results were compared to ISO 17225-6:2021 [14], which provides an initial classification for biofuels based on their origin and source. This standard distinguishes between various groups: woody, herbaceous, fruit, and aquatic biomass. Among the specifications for densified products of non-woody pellets, two options—A and B—were delineated, ranging from the most to the least demanding. These specifications are detailed in Table 2.

Table 2 Main specifications of pellets quality

	A	B
Diameter (mm)	6 ± 1	6 ± 1
Length (mm)	3.15–40	3.15–40
Moisture (%)	≤12	≤15
Durability (%)	≥97.5	≥96
Density (kg/m^3)	≥600	≥550

To determine the possible influence of the test parameters, i.e. moisture and mass flow, on the characteristics of the pellets produced, the Pearson correlation coefficient was calculated. Pearson's correlation coefficient (r) is a common way of measuring a linear correlation between any two variables, as it is considered the first formal measure of correlation and is still the most widely used method [15]. To perform this calculation, the R software version 4.3.2 was used in combination with the R Commander package.

3 Results and Discussion

The main results obtained in this research, as well as a discussion of them, are included below.

3.1 Characterisation of the Raw Material

The results of proximate and ultimate analyses, as well as the moisture, density and higher heating value, are shown in Table 3.

Comparing the results obtained with the standards regarding the characteristics of non-woody pellets, ISO 17225-6:2021 [14], it was observed that the moisture content was lower than the standard requirement, and pre-drying of the raw material is unnecessary. Taking into consideration said standard, and based on the results obtained, compliance with the ash content requirements (<6%) and the higher heating value (> 14.5 MJ/kg) was observed, making this residue optimal to produce Type A pellets, which have higher demands. The N and S content requirements for Type A pellets were also met.

Table 3 Proximate and ultimate analyses and HHV

Moisture (wb*)	Density (wb*)	C (db*)	H (db*)	N (db*)	S (db*)	Ash (db*)	Volatile matter (db*)	Fixed carbon (db*)	HHV (db*)
10.29%	118.26 kg/m^3	43.89%	7.10%	0.52%	0.04%	1.95%	82.2%	15.85%	17.18 MJ/kg

*HHV: higher heating value, wb: web basis, db: dry basis

3.2 Pelletisation Test Results

Table 4 shows the results obtained for each of the tests carried out, marking with the letter A those tests that have complied with the specifications for Type A pellets, for each specific characteristic. Likewise, the tests that meet the requirements for Type B pellets are marked with the letter B. Finally, those that do not meet the minimum requirements to produce non-woody pellets, according to ISO 17225-6:2021, have been marked with an X.

Table 4 Characteristics of corn residue pellets

Test number	Diameter (mm)	Length (mm)	Moisture (%)	Durability (%)	Density (kg/m^3)	Pellets type
A1	A	A	A	X	A	**X**
A2	A	A	A	B	A	**B**
A3	A	A	A	B	A	**B**
A4	A	A	A	A	A	**A**
A5	A	A	A	A	A	**A**
A6	A	A	A	A	A	**A**
A7	A	A	A	B	A	**B**
B1	A	A	A	X	A	**X**
B2	A	A	A	A	A	**A**
B3	A	A	A	A	A	**A**
B4	A	A	A	A	A	**A**
B5	A	A	A	A	A	**A**
B6	A	A	A	A	A	**A**
B7	A	A	A	B	A	**B**
C1	A	A	A	X	A	**X**
C2	A	A	A	B	A	**B**
C3	A	A	A	B	A	**B**
C4	A	A	A	B	A	**B**
C5	A	A	A	B	A	**B**
C6	A	A	A	B	A	**B**
C7	A	A	A	B	A	**B**
D1	A	A	A	X	A	**X**
D2	A	A	A	X	A	**X**
D3	A	A	A	B	A	**B**
D4	A	A	A	X	A	**X**
D5	A	A	A	B	A	**B**
D6	A	A	A	B	A	**B**
D7	A	A	A	B	A	**B**

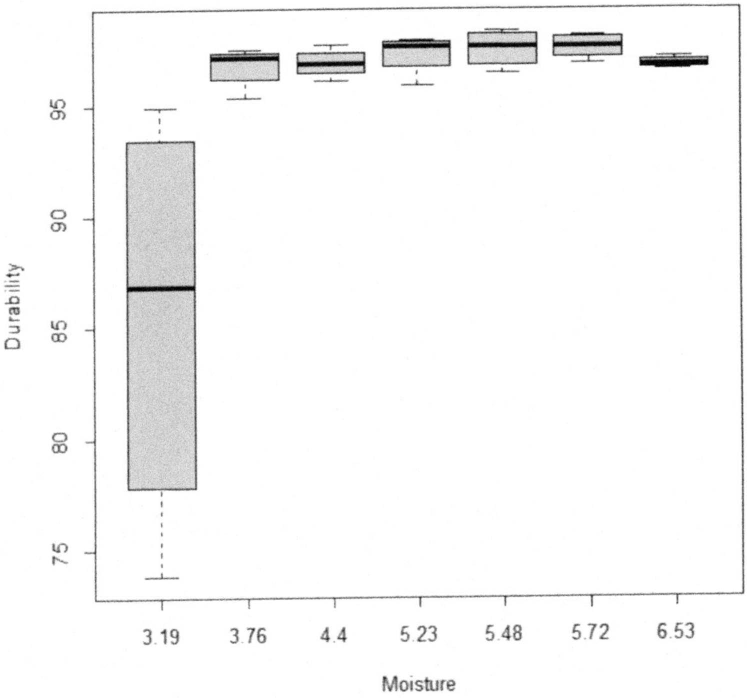

Fig. 2 Moisture—Durability box plot

As can be seen in the table, in all cases the corn pellets meet the diameter, length, moisture and density requirements of type A pellets, with durability being the limiting parameter for this type of pellets.

It is observed that the durability was lower than the minimum established when the water input to the process was 3.19 g/min, regardless of the corn flow input at the pelletiser inlet, as can be seen in Fig. 2.

This finding suggests that the significance of the pelleting process extends beyond the initial moisture content, also considering the importance of water supply throughout the pelleting operation.

It was also observed that only class A pellets were obtained for mass flows of 302 and 462 g/min. This can be seen in Fig. 3, which shows the box plot of durability versus inlet mass flow, for durability values above 96%.

In view of the results obtained, the best test conditions were those with a mass flow of between 302 and 462 g/min combined with a water input of between 5.23 and 5.72 g/min. Higher mass flow rates or water additions do not improve the quality of the pellets. This occurs because mass flow rates below 302 g/min do not provide sufficient material for the necessary compaction to take place in pelletizing, and the pellets do not have sufficient strength and durability is reduced, while flow rates above 462 cause the equipment to collapse and overheating of the raw material occurs which impairs durability. In relation to the water supply, flow rates lower than 5.23

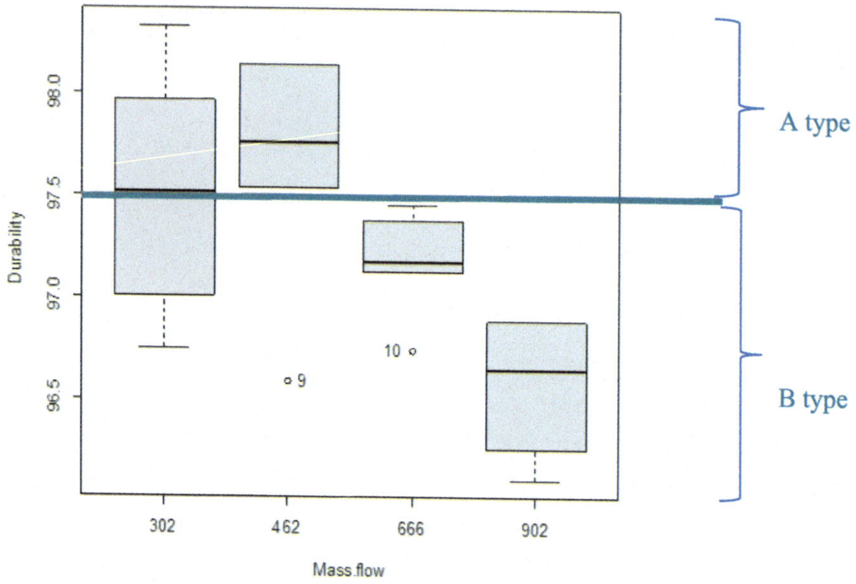

Fig. 3 Mass flow—Durability box plot (durability > 96%)

g/min do not allow pelletizing, as the temperature of the flat matrix of the pelletizer increases greatly, while with excessive water supply, above 5.72 g/min, the necessary compaction of the pellets does not occur and the pellets break.

To determine whether there is a correlation between the durability of produced pellets and the initial conditions, the Pearson correlation coefficient has been calculated, yielding the following correlation matrix (Table 5).

From the results obtained, it is worth highlighting the influence that the water provided in the process has on the durability of the pellets, so that when designing a pelletizing process it is not only necessary to consider the starting moisture of the raw material but also the water provided during the process.

Table 5 Pearson correlation coefficient

	Durability	Mass flow	Moisture	R
Durability	1.000	0.141	0.519	0.171
Mass flow	0.141	1.000	0.000	0.823
Moisture	0.519	0.000	1.000	0.464
R	0.171	0.823	0.464	1.000

4 Conclusions

The principal findings of the conducted studies are summarized as follows:

- Corn residues, particularly corn cob waste, were confirmed as optimal materials for the production of non-woody pellets.
- Characterization of the raw material demonstrated compliance with reference standards regarding moisture, nitrogen (N) and sulphur (S) content, ash content, and higher heating value.
- Pelleting tests consistently met the requirements for Type A pellets, according to ISO standards, with respect to diameter, length, moisture, and density across all cases.
- Durability testing yielded optimal results for mass flow rates ranging from 302 to 462 g/min and water input between 5.23 and 5.72 g/min.
- Of the initial parameters studied, it was determined that the water added to the process had the greatest influence on the durability of the pellets.

Acknowledgements This research was funded by "Operational Programme ERDF Extremadura 2021-2027" in the framework of Action 1A1103 "Development of Scientific Research Capacity, Technological development and Innovation in CICYTEX PO 21-27" SMARTCIRCLE Project.

References

1. Food and Agriculture Organization of the United Nations. FAOSTAT (2020). http://fao.org/faostat/en/#data. Accessed 1 Sept 2020
2. Miranda MT, García-Mateos R, Arranz JI, Sepúlveda FJ, Romero P, Botet-Jiménez A (2021) Selective use of corn crop residues: energy viability. Appl Sci (Switzerland) 11. https://doi.org/10.3390/app11073284
3. ISO 16948:2015. Solid biofuels. Determination of total content of carbon, hydrogen and nitrogen (2015)
4. ISO 16994:2016. Solid biofuels. Determination of content of sulfur and chlorine (2017)
5. ISO 18134-2:2017: Solid biofuels. Determination of moisture content. Oven dry method. Part 2: Total moisture. Simplified method (2017)
6. ISO 18122:2022. Solid biofuels. Determination of ash content (2022)
7. ISO 18123:2023 Solid biofuels—Determination of volatile matter (2023)
8. ISO 18125:2017. Solid biofuels. Determination of calorific value (2017)
9. Miranda MT, Sepúlveda FJ, Arranz JI, Montero I, Rojas CV (2018) Physical-energy characterization of microalgae Scenedesmus and experimental pellets. Fuel. https://doi.org/10.1016/j.fuel.2018.03.097
10. Mani S, Tabil LG, Sokhansanj S (2006) Effects of compressive force, particle size and moisture content on mechanical properties of biomass pellets from grasses. Biomass Bioenergy 30:648–654. https://doi.org/10.1016/j.biombioe.2005.01.004
11. Pradhan P, Mahajani SM, Arora A (2018) Production and utilization of fuel pellets from biomass: a review. Fuel Process Technol 181:215–232. https://doi.org/10.1016/j.fuproc.2018.09.021
12. ISO 17831-1:2015. Solid biofuels. Determination of mechanical durability of pellets and briquettes. Part 1: Pellets (2015)

13. ISO 17829:2015. Solid biofuels. Determination of length and diameter of pellets (2015)
14. ISO 17225-6:2021. Solid biofuels. Fuel specifications and classes. Part 6: Graded non-woody pellets (2022)
15. Chen Y, Wang Z, Lin S, Qin Y, Huang X (2023) A review on biomass thermal-oxidative decomposition data and machine learning prediction of thermal analysis. Clean Mater 9. https://doi.org/10.1016/j.clema.2023.100206

Assessing the Inclusiveness of the Draft Updated National Energy and Climate Plans 2030 of Belgium and Portugal Through a Hard-to-Reach Energy Users Lens

Miguel Macias Sequeira and João Pedro Gouveia

Abstract The European Union's energy strategy vows to strive for a fair transition where no one is left behind. Hard-to-reach energy users encompass residential and non-residential groups generally neglected by energy policies. National Energy and Climate Plans (NECP) for 2030 are a key instrument currently being updated by the Member States. In this context, we aim to put the commitment to a just transition to the test by applying a hard-to-reach energy users framework to the draft updated NECPs of Belgium and Portugal. While vulnerable households are mentioned in the plans and tailored and targeted measures are proposed, specific profiles such as rural households, migrants, people with ill-health, and other marginalized groups seem to be overlooked, particularly in the Portuguese case. The need to address excessive energy consumption of high-income households has been the focus of recent research. Still, Belgium's NECP scantily mentions this group, which is absent in Portugal's NECP. Specific schemes for the renovation of rented homes are presented in both NECPs but with a much greater level of detail in Belgium's one. Small and medium enterprises represent most businesses and are mentioned in the NECPs, while a much smaller focus is given to micro-enterprises. Commercial subsectors seem to be adequately mentioned, although activities such as food services do not attract much attention. We conclude by arguing that, even if some promising examples can be found, policymakers can do more to ensure the engagement of hard-to-reach groups in the energy transition.

Keywords Energy transition · Energy policy · Vulnerable households · NECP

M. M. Sequeira (✉) · J. P. Gouveia
CENSE—Center for Environmental and Sustainability Research & CHANGE—Global Change and Sustainability Institute, NOVA School of Science and Technology, NOVA University of Lisbon, Caparica, Portugal
e-mail: m.sequeira@campus.fct.unl.pt

J. P. Gouveia
e-mail: jplg@fct.unl.pt

P. S. Brito et al. (eds.), *Insights into Energy Trends*, Green Energy and Technology,
https://doi.org/10.1007/978-3-031-83811-8_6

1 Introduction

The European Union (EU) has vowed to be a leader in climate change mitigation, foreseeing a net reduction of greenhouse gas emissions of at least 55% in 2030 compared to the 1990 baseline [1]. This goal is to be achieved by raising the share of renewable energy in the EU's overall energy consumption to 42.5% and improving energy efficiency to reduce final energy consumption at the EU level by 11.7%. Recognizing the different responsibilities and needs of specific population groups, the European Green Deal emphasizes a just transition where no one is left behind [2].

In this context, the concept of hard-to-reach (HTR) energy users has been gaining attention from researchers, policymakers, and practitioners, broadly described as residential and non-residential energy users that have so far been neglected by energy policies and interventions [3]. Five major HTR groups are identified in the literature, namely vulnerable households, high-income households, tenants and landlords, small and medium enterprises, and commercial subsector. A few authors have endeavored to establish more precise residential and non-residential HTR profiles [4, 5].

Due to multiple and persistent barriers, the involvement of HTR groups in the energy transition has been slow, hindering both the success and fairness of this societal transformation. A recent stream of literature has argued for targeted and tailored energy policies that recognize the characteristics and needs of HTR households and businesses and detail concrete approaches to surpass these challenges [6].

A key research question with practical policy-oriented implications is to assess if current energy policies are being inclusive towards HTR energy users and to what extent. In this work, we explore the draft updated National Energy and Climate Plans (NECP) for 2030 of EU Member States Belgium and Portugal for mentions of HTR groups regarding just energy transitions and targeted and tailored measures. By applying an HTR energy users framework to these policy documents, which are currently being revised by Member States, we aim to identify forgotten groups at risk of being left behind by the EU's energy transition.

2 Methods

This section describes the methods applied in this research, namely the systematic approach used to assess policy instruments - i.e., the NECPs for 2030 of Belgium and Portugal - by applying a HTR energy users theoretical framework.

2.1 National Energy and Climate Plans 2030 of Belgium and Portugal

In the EU, the NECPs outline the Member States' 10-year strategic plans for the energy and climate area, addressing five key dimensions, namely decarbonization, energy efficiency, energy security, internal energy market, and research, innovation, and competitiveness [7]. Member States were required to submit their NECPs by December 2019, taking into consideration the European Commission's assessment of their draft version. A progress report on the implementation of the NECPs is due every two years.

In 2023, Member States were mandated to update their NECPs in line with the EU's revised energy and climate targets for 2030. The draft versions were due to be submitted in June 2023, and the European Commission published its assessment and issued recommendations to raise ambition in December 2023. Member States must submit their final NECPs by June 2024. For an improved development and implementation of the NECPs, Member States were required to consult stakeholders in the drafting and finalization process.

Belgium and Portugal submitted their draft updated NECPs per the European Commission's timeline [8, 9]. These countries have similar population levels, but present markedly different climates and socio-economic conditions (€117 and €83 GDP per capita for Belgium and Portugal, respectively). Furthermore, Portugal and Belgium have distinct primary energy mixes (64% and 75% of fossil fuels in the energy mix for Portugal and Belgium, respectively), greenhouse gas emissions levels (86 and 35 Mt CO_2, for Belgium and Portugal, respectively), and energy poverty problems (6% and 21% of the population unable to heat their homes in Belgium and Portugal, respectively). Thus, these two contrasting countries were selected for methodological testing purposes, and future work will analyze the NECPs of all Member States. The NECP of Belgium consists of 758 pages, while the Portuguese one has only 238 pages. These countries are in the revision process of their NECPs, making the current analysis valuable and timely.

2.2 Hard-to-Reach Energy Users' Framework

In this work, we use the HTR energy users framework developed by [4] for the residential sector and by [5] for the non-residential sector (Fig. 1). These authors built on the existing scientific literature on major HTR groups to further define specific HTR profiles. Furthermore, they operationalized it using existing standardized statistical data to gauge the significance of these audiences at the EU and Member State scale. This framework is now applied to assess if energy policies are (i) recognizing the relevance of HTR groups for just energy transitions and (ii) defining targeted and tailored measures that meet the needs of specific HTR profiles.

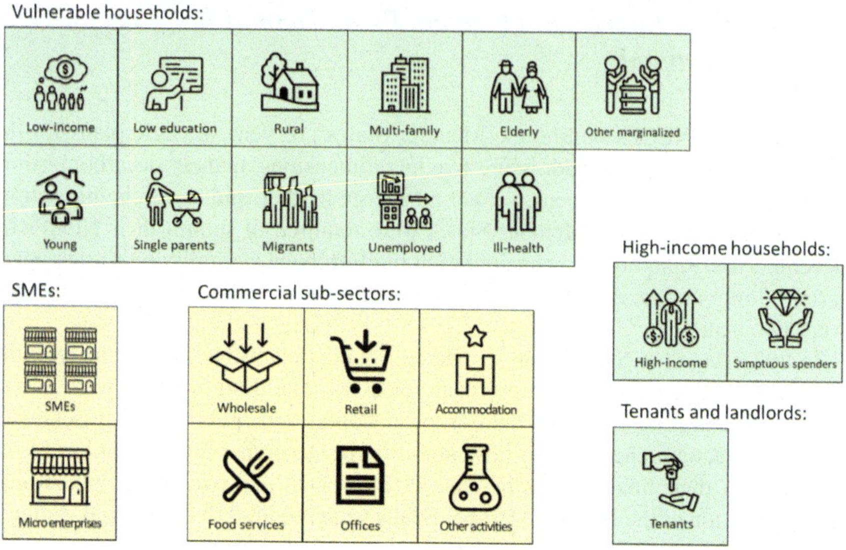

Fig. 1 Hard-to-reach energy users framework (adapted from [4, 5])

2.3 *Assessing Energy Policy Through a Hard-to-Reach Lens*

Considering the HTR energy users framework, the NECPs of Belgium and Portugal were scanned for mentions of HTR groups. To perform this task, representative keywords were selected for each HTR profile, as presented in Tables 1 and 2 for the residential and non-residential sectors, respectively. Only the successful keywords are shown since a wide range of other keywords was tested during this process. Furthermore, the sections where the keywords were found were read in detail to understand the context in which HTR groups were mentioned.

The goal was to identify the approaches to recognizing and engaging these HTR groups in the energy transition as being "energy users" and not in other roles. Thus, several topics were deemed out-of-scope for this assessment (e.g., international climate finance, vulnerable nature, climate adaptation, and green jobs, among others).

Figure 2 showcases the matrix used to apply the HTR energy users framework to the draft updated NECPs. Each cell represents the assessment of one HTR profile for a given country NECP. The share of the HTR profile as a percentage of the population or of the number of enterprises for the residential and non-residential sectors, respectively, was retrieved from [4, 5]. The number of mentions in the draft updated NECP only encompasses the topics that are considered in the scope of this work, namely just energy transitions and energy poverty and several types of targeted and tailored measures.

Finally, each cell is given a color that considers both the number of mentions and the representativeness of the HTR profile in the total population or number of

Table 1 Keywords to scan the NECPs for mentions of HTR profiles (residential sector)

HTR group	HTR profile	Keywords
Vulnerable households	Low-income	Low/lower/lowest income, poor socio-economic
	Low education	Low skilled, info-exclusion
	Rural	Rural, remote
	Multi-family	Apartment, multi-dwelling, condominium
	Elderly	Elderly, older, aged 65 and over
	Young	Young adults/people
	Single parents	Single parents/households/mothers, divorced
	Migrants	Migrants, intercultural
	Unemployed	Unemployed, not currently working
	Ill-health	Poor health, disabilities, illness, health problems
	Other marginalized	–
High-income households	High-income	Higher/highest income
	Sumptuous spenders	–
Tenants and landlords	Tenants	Tenant, rental stock

Table 2 Keywords to scan the NECPs for mentions of HTR profiles (non-residential sector)

HTR group	HTR profile	Keywords
Small and medium enterprises	SMEs	SME, small and medium-sized enterprises
	Micro-enterprises	Micro-enterprises, small business
Commercial sub-sectors	Wholesale	Depot, distribution sector
	Retail	Retail, shop, shopping, trade, commerce
	Accommodation	Horeca, hotel, tourism sector
	Food services	Horeca
	Offices	Office
	Other activities	Sports, care, laundry, hospital

enterprises of the country. This is done by dividing the number of mentions by the share of the population or enterprises; if the result is larger than 0 and smaller than 0.5, a yellow color is attributed; if it is equal or greater than 0.5, a green color is attributed. The cell is shown as red if the HTR profile is not mentioned in the NECP.

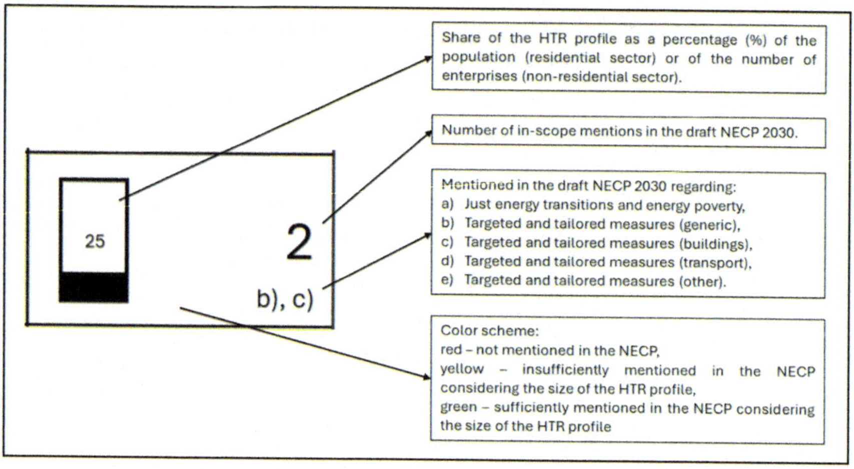

Fig. 2 Assessment matrix for the application of the HTR framework to the NECPs

3 Results and Discussion

This section describes the main outputs from the application of the HTR energy users framework to the NECPs for 2030 of Belgium and Portugal, for the residential and non-residential sectors.

3.1 Residential Sector

For the residential sector, the assessment of the draft NECPs of Belgium and Portugal using the HTR framework is shown in Fig. 3. The NECP of Belgium seems to perform better than the Portuguese, partially because of its greater level of detail.

The low-income HTR profile is mentioned in both NECPs, although insufficiently in the Portuguese case, given the sheer size of this target group. For Belgium, the NECP establishes the need for a just energy transition that mitigates energy poverty and shares the benefits of clean energy. Targeted measures include specialized support and dedicated financing for low-income households to renovate their houses and install solar panels. The Portuguese NECP only mentions low-income households two times. Nevertheless, it includes targeted measures describing on-the-ground support and dedicated financing of home renovation interventions.

Even though a relevant share of each country's population has low education levels and might potentially be more challenging to reach with energy interventions due to energy and digital illiteracy, this is not adequately considered in the NECPs. The Portuguese NECP mentions the need to provide support to people in a situation of info-exclusion. At the same time, the Belgian NECP argues for the promotion of the social inclusion of low-skilled people.

HTR profile \ Member State	Belgium	Portugal
Low income	29 — 24 a), b), c)	25 — 2 b), c)
Low education	13 — 1 b)	24 — 1 b)
Rural	15 — 4 a), d)	23 — 1 e)
Multi-family	22 — 25 a), c)	47 — 3 c), d)
Elderly	18 — 13 a), c), d)	17 — 0
Young	12 — 14 a), b), c), d)	5 — 4 a), b), d)
Single-parents	6 — 9 a), c)	4 — 0

HTR profile \ Member State	Belgium	Portugal
Migrants	18 — 2 b), c)	12 — 0
Unemployed	2 — 2 b)	3 — 0
Ill-health	22 — 5 a), c), d)	39 — 1 c)
Other marginalized	? — 0	? — 0
High-income	11 — 2 a), c)	19 — 0
Sumptuous spenders	? — 0	? — 0
Tenants	28 — 25 a), c)	22 — 1 c)

Fig. 3 Mentions of residential HTR profiles in the NECPs of Belgium and Portugal

Regarding rural areas, most measures in the NECPs focus on stimulating socio-economic development and do not specifically mention rural households. These can be considered HTR due to geographical isolation, lack of available services, and use of unregulated energy carriers. The Belgian NECP states the need to involve rural people in energy policymaking while providing targeted measures for households' transport energy use. The Portuguese NECP only presents one measure focused on improving grid access to remote populations.

Interventions in multi-family buildings are hindered by ownership and organizational challenges. Belgium's NECP recognizes this well, providing specific financial and technical support schemes for renovating apartment buildings, including through the involvement of co-owners associations in the process. While in Portugal, almost half of the population lives in apartments, this target group is scantily mentioned, stating only the need to develop reference buildings and a single measure for installing electric vehicle charging stations in condominiums.

Elderly people living alone or as a couple can be particularly susceptible to energy poverty while being harder-to-reach with energy interventions. The Belgian NECP recognizes this exacerbated vulnerability and designs tailored measures such as loans for renovating homes (older homeowners typically do not have access to traditional bank credits) and lower prices for public transport fares. This target group is not mentioned in Portugal's NECP, even if it represents a significant population share.

Both NECPs adequately mention young people. Young adults living independently (the indicator depicted in the table) can be considered HTR due to unstable and transient housing and employment. In Portugal, this group is much less significant than in Belgium. Both NECPs mention the need to engage with youth for a just energy transition while providing measures to improve energy literacy, promote behavioral change, and increase public transportation use.

The increased vulnerability to energy poverty in single-parent households is well described in Belgium's NECP, which also frames the gender dimension of this issue. Migrants can be considered HTR due to language barriers and lack of skills to navigate the energy market in the host country. In Belgium, migrants account for almost one-fifth of the total population, and their active participation in the energy transition is vital. However, these groups are only mentioned two times in the NECP regarding general measures to promote social inclusion and foster the engagement of intercultural stakeholders. Unemployment rates have been dropping around Europe, and Belgium's NECP includes only a few mentions of this group. In contrast, Portugal's NECP does not recognize the need to engage with households matching the profiles of single parents, migrants, and the unemployed.

Research has uncovered a bi-directional relationship between energy poverty and ill-health [10]. This is recognized in both NECPs, describing measures to protect consumers with health problems and proposing measures for enhanced access to public transport. Other marginalized groups, such as the homeless, ethnic minorities, indigenous people, and criminalized populations, among others, have historically been left out of energy policies; the assessed NECPs are no exception to this trend.

While energy poverty has been identified as a key issue for just energy transitions, attention is also growing to the opposite side of the social ladder and issues of excess energy use [11]. The Belgian NECP timidly recognizes this issue, ascertaining different responsibilities and outcomes for higher and lower income households and foreseeing stricter measures for high-income households to renovate their buildings. Portugal's NECP does not mention high-income households. Not mentioned in both NECPs is the narrower high-income group described as sumptuous spenders whose energy use largely surpasses the average household due to luxury consumption.

Finally, the private rental sector can be considered one of the harder-to-reach due to split incentives between landlords and tenants and other barriers. With more than one-quarter of households living in rented homes, Belgium's NECP recognizes the increased risk of energy poverty for tenants and defines measures to facilitate renewable energy sharing and energy renovation while seeking to protect tenants from rent increases. While Portugal's NECP mentions the need to rehabilitate social housing, it only mentions tenants once regarding a program to renovate privately owned housing, which is subsequently made available at affordable rent.

3.2 *Non-residential Sector*

For non-residential HTR profiles, the assessment of the NECPs of Belgium and Portugal using the HTR energy users framework is shown in Fig. 4.

Small and medium enterprises represent the overwhelming majority of enterprises in Belgium and Portugal, where barriers such as lack of knowledge, financing difficulties, and other priorities may hinder the uptake of energy interventions; these challenges are particularly severe in micro-businesses. Given the sheer size of these groups, it can be considered that they could be more extensively mentioned in the NECPs, particularly in the Portuguese one. Nevertheless, the plans frame the need to support smaller businesses in the energy transition while providing a few targeted and tailored measures, such as carrying out energy audits (with or without support depending on the size of the enterprise), implementing interventions with short payback times (mandatory), installing renewable energy systems, establishing awards for outstanding enterprises, withdrawing the use of F-gases, among others.

Fig. 4 Mentions of non-residential HTR profiles in the NECPs of Belgium and Portugal

Member State / HTR profile	Belgium	Portugal
Small and medium enterprises	99,9 33 a), b), c), d), e)	99,9 6 b)
Micro-enterprises	96 2 d), e)	95 1 a)
Wholesale	7 5 d), e)	6 3 b), d)
Retail	11 10 c), e)	14 6 a), c)
Accommodation	1 1 c)	4 7 a), b), c), d)
Food services	7 1 c)	8 0
Offices	20 8 c), d)	27 3 a), c)
Other services	24 25 c)	15 5 a), c)

Commercial sub-sectors include a heterogeneous range of activities and services often neglected in energy policies. For the Belgian and Portuguese NECPs, this does not seem to be the case. A set of targeted measures is in place for the improvement of the energy performance of wholesale, such as capacity building and electrification of transport, and of retail, such as raising awareness on the need to close doors in heated or cooled establishments, replacing refrigeration equipment with more efficient ones, and promoting water efficiency. The need to improve energy performance in the tourism sector is extensively mentioned in the Portuguese NECP, with measures targeting buildings and transport, reflecting the sector's importance in the national context. Food services, such as restaurants and bars, are only mentioned in the Belgian NECP, where they are aggregated with the accommodation sector. The need to renovate office buildings is mentioned in both NECPs. The other activities profile encompasses sports, health, laundry, and other services, some of which are mentioned in the NECPs, including, for instance, through targeted measures for sports buildings in the Belgian plan.

4 Conclusions

In this research, we have performed a cross-evaluation of the NECPs of Belgium and Portugal by applying an HTR energy users framework to scan for specific mentions of HTR households and businesses and understand the extent of their inclusion. NECPs are the key energy and climate policy instruments for the EU Member States, outlining the country's strategy until 2030. These are currently being updated, and it is critical to provide science-based recommendations that can contribute to their revision process. The following insights can be summarized:

(i) Vulnerable households and energy poverty are mentioned in Belgium's and Portugal's NECPs, but specific HTR profiles are often overlooked. For both countries, these include people with low education levels, rural households, migrants, people with ill health, and other marginalized groups. The Portuguese NECP does not adequately mention other relevant groups, such as elderly people, multi-family buildings, single parents, and the unemployed. Recognition of these HTR groups can be improved, and more targeted and tailored measures can be outlined.

(ii) High-income households are timidly mentioned in Belgium's NECP and completely overlooked in the Portuguese one. In line with recent research, this implies that more attention needs to be drawn to the issue of excessive energy consumption, mainly due to luxury consumption, if a just energy transition is to be enabled.

(iii) Due to split incentives and other challenges, the rented building stock has historically been considered harder to reach. A relevant share of Belgium's population lives in a rented home. The NECP outlines targeted and tailored measures so tenants and landlords can find ways to improve energy performance. The Portuguese NECP does not adequately mention this issue.

(iv) Small and medium enterprises represent over 99% of enterprises in both countries. The NECPs provide targeted measures to protect businesses from high energy prices and support them during the energy transition. Nevertheless, micro-enterprises where the challenges are particularly severe are scarcely mentioned.

(v) The commercial/services sector is highly heterogeneous, with a wide range of activities and energy use patterns being considered HTR. Both NECPs adequately mention different subsectors within this group, establishing targeted measures. Specific subsectors, such as food services, seem to attract less attention.

While promising examples of recognition of HTR groups and targeted and tailored measures can be found, we argue that policymakers can do much more to ensure that energy policies really promote an ambitious, fast, and just energy transition that leaves no one behind.

Acknowledgements CENSE, NOVA FCT is supported by the Portuguese Foundation for Science and Technology (FCT) through project UIDB/04085/2020. Miguel Macias Sequeira's PhD scholarship is funded by FCT (https://doi.org/10.54499/2020.04774.BD).

References

1. European Commission (2021) 'Fit for 55': delivering the EU's 2030 climate target on the way to climate neutrality. Brussels, 14.7.2021. COM (2021) 550 final
2. European Commission (2019) The European Green Deal. Brussels, 11.12.2019. COM (2019) 640 final
3. Rotmann S, Mundaca L, Castaño-Rosa R, O'Sullivan K, Ambrose A, Marchand R, Chester M, Karlin B, Ashby K, Butler D, Chambers J (2020) Hard-to-reach energy users: a literature review. User-Centred Energy Systems TCP—HTR Task. ISBN: 978-0-473-64983-8
4. Sequeira MM, Gouveia JP, Melo JJ (2024) (Dis)comfortably numb in energy transitions: gauging residential hard-to-reach energy users in the European Union. Energy Res Soc Sci 115:103612. https://doi.org/10.1016/j.erss.2024.103612
5. Sequeira MM, Gouveia JP (2023) Non-residential hard-to-reach energy users: a vast audience forgotten by the energy transition? In: 8th meeting on energy and environmental economics, Aveiro University, Portugal. https://doi.org/10.48528/v3b3-fk78
6. Mundaca L, Rotmann S, Ashby K, Karlin B, Butler D, Sequeira MM, Gouveia JP, Palma P, Realini A, Maggiore S, Feenstra M (2023) Hard-to-reach energy users: an ex-post cross-country assessment of behavioural-oriented interventions. Energy Res Soc Sci 104:103205. https://doi.org/10.1016/j.erss.2023.103205
7. European Commission (2024) National energy and climate plans: EU countries' 10-year national energy and climate plans for 2021–2030. https://commission.europa.eu/energy-climate-change-environment/implementation-eu-countries/energy-and-climate-governance-and-reporting/national-energy-and-climate-plans_en
8. Government of Belgium (2023) Draft update of the Belgian National Energy and Climate Plan. https://commission.europa.eu/publications/belgium-draft-updated-necp-2021-2030_en

9. Government of Portugal (2023) Draft National Energy and Climate Plan 2021–2030 (NECP 2030): Update/Revision. https://commission.europa.eu/publications/portugal-draft-updated-necp-2021-2030_en
10. Ivanova D, Middlemiss L (2021) Characterizing the energy use of disabled people in the European Union towards inclusion in the energy transition. Nat Energy 6:1188–1197. https://doi.org/10.1038/s41560-021-00932-4
11. Gossling S, Humpe A (2023) Millionaire spending incompatible with 1.5 °C ambitions. Clean Prod Lett 4:100027. https://doi.org/10.1016/j.clpl.2022.100027

An Assessment of Wind Resources with Micro-Generation Applied on Farmland in Inland Portugal

Vanderlei Cruz, Natália Roque, António Canatário Duarte, and António Cerejo

Abstract Since immemorial time, wind energy has been exploited. Taking advantage of the progress of technology, it has become possible to produce electricity through wind turbines in windy locations, even in the most remote places. In various parts of the world, especially in areas with favorable conditions, it has become evident that such alternative energy corresponds to an extremely interesting choice. Portugal is no exception and is increasingly committed to wind energy, whose participation in energy production has stood out over time. In 2021, 26% of Portugal's electricity came from wind, which positions the country as one of the leaders in using this renewable resource. This project focuses on analyzing the direction and intensity of winds in the Beira Baixa region, expanding this analysis to the possibility of using wind turbines in mini and micro-generation next to the rural population and the villages, but not neglecting the possibilities of integration with other renewable energy sources, through hybrid systems, such as solar and hydro.

Keywords Energy sustainability · Portugal and wind energy · Renewable sources integration · Sustainable agriculture · Wind energy

1 Introduction

In the context of the energy transition, the wind sector and photovoltaics have been the sectors with the strongest growth over the last 25 years [1]. However, this important form of energy production presents problems of predictability. Various mathematical and meteorological forecasting models are used for wind forecasting, making estimating the power generated possible [2–10]. Even so, it is common for forecasts to fail, leading to deviations in the power delivered from that previously agreed. This

V. Cruz · N. Roque (✉) · A. C. Duarte · A. Cerejo
Instituto Politécnico de Castelo Branco (IPCB), Castelo Branco, Portugal
e-mail: nroque@ipcb.pt

N. Roque · A. C. Duarte
CERNAS-IPCB Research Centre for Natural Resources, Environment and Society, Polytechnic Institute of Castelo Branco, Castelo Branco, Portugal

P. S. Brito et al. (eds.), *Insights into Energy Trends*, Green Energy and Technology,
https://doi.org/10.1007/978-3-031-83811-8_7

missing power will be replaced by other units in reserve to meet demand, but this has increased costs for producers who default on power.

An adequate characterization of the wind tends to give a more accurate estimate of the power generated, which allows the decision-maker to make a more precise economic feasibility analysis.

In a rural environment, and given the high costs involved in connecting to a distant electricity grid, it is not always economically viable for farms to access electricity. In this context, isolated electricity micro-production systems have emerged as a solution to be considered when supplying electricity. The most efficient systems are hybrid, comprising a wind farm and a photovoltaic farm.

As photovoltaic production is highly predictable, this article presents a wind characterization study to implement a wind or hybrid system on farms.

1.1 Objectives

This study aims to analyze the wind potential in the Beira Baixa region, especially concerning the possibilities for residential and rural self-consumption. Focusing on this, it is possible to assess the wind conditions in Beira Baixa (speed and direction) to determine the feasibility of small-scale production. Digital modeling allows for locating regions within the study area that offer the most suitable conditions for installing small-scale wind turbines for rural and/or domestic use.

2 Material and Methods

In the material and methods section, we present the study area, assess the basic data available, its standardization and the methodology for the potential wind resource assessment.

2.1 Study Area

The Intermunicipal Community of Beira Baixa (CIMBB), which at the time of this study included the municipalities of Penamacor, Idanha-a-Nova, Castelo Branco, Oleiros, Proença-a-Nova and Vila Velha de Rodão, is a region with essential water resources, which can provide profitable use of the land, and forest areas that are usually part of natural landscapes of great ecological value. It is also a region where farming and cattle-raising activities on small or large properties are significant, particularly arable crops for animal feed, such as peach orchards. The small-scale generation or co-generation of electricity from renewable sources, such as wind, could be an interesting alternative for these areas where livestock farming takes place.

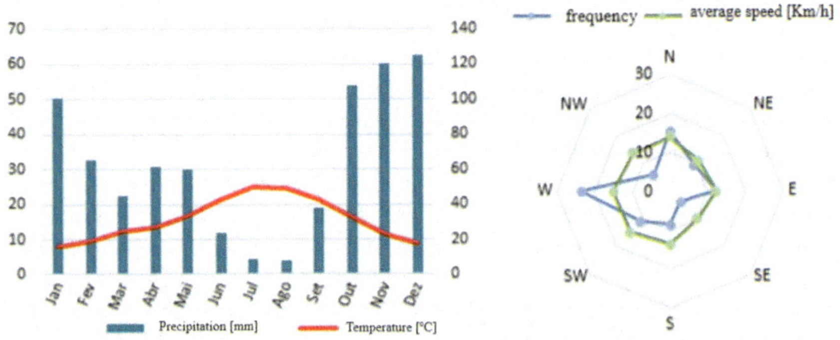

Fig. 1 Ombrothermal diagram and wind roses for Castelo Branco weather station

The climate of the CIMBB is not specified by seasonal designations but rather by distinct climatic intervals, retrieved from IPMA [11, 12]. These climatic phases immediately impact the wind's transformation throughout the year (Fig. 1).

The CIMBB has large areas of fertile land occupied by farms with a variety of crops, which are an integral part of the local economy. This ties in closely with the collection of wind energy, as turbines can be skillfully installed to produce electricity for use on these properties, which includes irrigation systems, heating greenhouses, animal farms, and other agricultural facilities [13].

Almost half of the region's land use is artificialized, the largest percentage of which is forest, scrubland, and surface water bodies. The municipalities of Idanha-a-Nova and Vila Velha de Rodão stand out for their higher percentage of agricultural land and, therefore, also for their consumption of electricity in agricultural activity (Fig. 2).

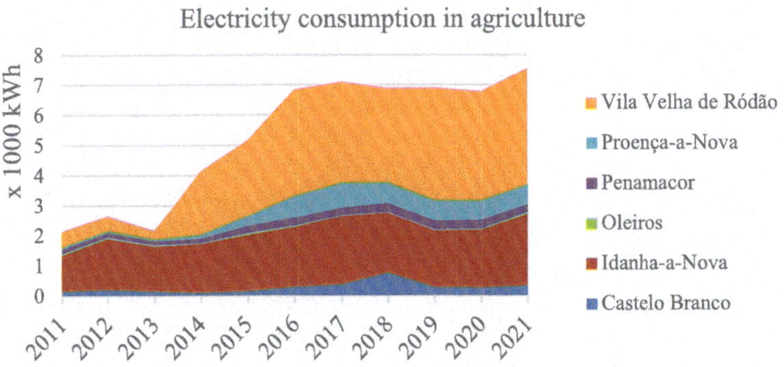

Fig. 2 Comparison of agricultural consumption 2011/2021

2.2 Availability and Wind Resource Parameters

The source of the data time series was the National Water Resources Information Service (SNIRH) [14], and the data was collected under the supervision of the Portuguese Environment Agency (APA).

The average hourly wind speed, measured at a height of 2 m from the ground surface, is necessary for extrapolating the speed for different wind turbine positioning heights (Eq. 1) [15] and subsequently calculating the energy that can be generated. The maximum hourly wind speed is also useful for knowing the characteristics of gusts (between 11.5 and 24.8 m/s) to define turbine safety procedures. The direction is important when deciding where to position the towers and wind turbines on the ground.

$$V_H = \vec{v} \cdot \left(\frac{H}{H_a}\right)^{\alpha} \tag{1}$$

where: $\boldsymbol{V_H}$ is the wind speed at the calculated height; $\vec{v}$ is the wind speed measured by the anemometer; $\boldsymbol{H}$ is the height at which the wind turbine is to be installed; $\boldsymbol{H_a}$ is the height of the anemometer; α is the wind shear exponent.

The exponent α is directly related to the roughness of the terrain where the wind turbines are installed. For hilly terrain or terrain with high turbulence, $\boldsymbol{\alpha} = 0.20$; relatively flat terrain, $\boldsymbol{\alpha} = 0.11$, although $\boldsymbol{\alpha} = 0.14$ is generally used. In this study,

Fig. 3 Location of the study area: dot (weather stations), polygon (CIMBB)

0.20 will be used as the limit between slightly rough and rough terrain, as it is the most representative situation in the study region.

For the calculations inherent in this study, 20 years of hourly data from 36 weather stations were used, 22 of which were within the study area and 14 outside of the study area, which were necessary for geostatistical operations (Fig. 3).

2.3 Wind Data Processing Methodologies

Spatial data was geoprocessed using ArcGIS (ESRI), Digital elevation model (DEM) layers, and land use data. The geostatistical methodology selected for processing wind speed data at different heights (2 m, 5 m, 10 m, and 20 m) was co-kriging between the point data and DEM (spatial resolution of 25 × 25 m).

This is an advanced geostatistical interpolation technique that exploits the spatial links between data points to extrapolate the characteristics of one piece of information, in this case, wind, with a second group of information, altitudes, considering not only wind data from weather stations but also the effect of terrain characteristics such as land use.

The methodological procedure comprises variogram analysis, which describes this spatial relationship. In this way, it is possible to ensure that the extrapolated data aligns with the local environment and the transformations that the landscape can generate, making the final data useful for assessing wind energy potential and enabling more adequate forecasts of wind action at different heights above the ground.

3 Results and Discussion

The results of the co-kriging show the general direction of the wind through the arrows and the maximum average wind speed intensity for the 20 years collected at the height of the anemometer (2 m). The wind velocity ranges from 9.5 [$m \cdot s^{-1}$] in the southeast of Idanha-a-Nova to 12.5 [$m \cdot s^{-1}$] at the confluence of Castelo Branco, Oleiros, and Proeça-a-Nova, Fig. 4. The predictive error of the model was analyzed, and it is presented in Appendix 1.

In the evaluation of the wind behavior in the climatic periods (Autumn, Winter, Spring, and Summer), the intensity of the average wind speed (AWS), which ranged from 0.96 to 1.2 [$m \cdot s^{-1}$], with a predominant direction from northeast to southwest, from the center of the region to the east, except winter where it showed more homogeneous characteristics of intensity (1.21 to 1.45 [$m \cdot s^{-1}$]), and direction from north to south throughout the region. Average speeds reach 1.46 to 1.70 [$m \cdot s^{-1}$] at the confluence of Castelo Branco, Oleiros, and Proeça-a-Nova, with the direction varying considerably according to the climatic period.

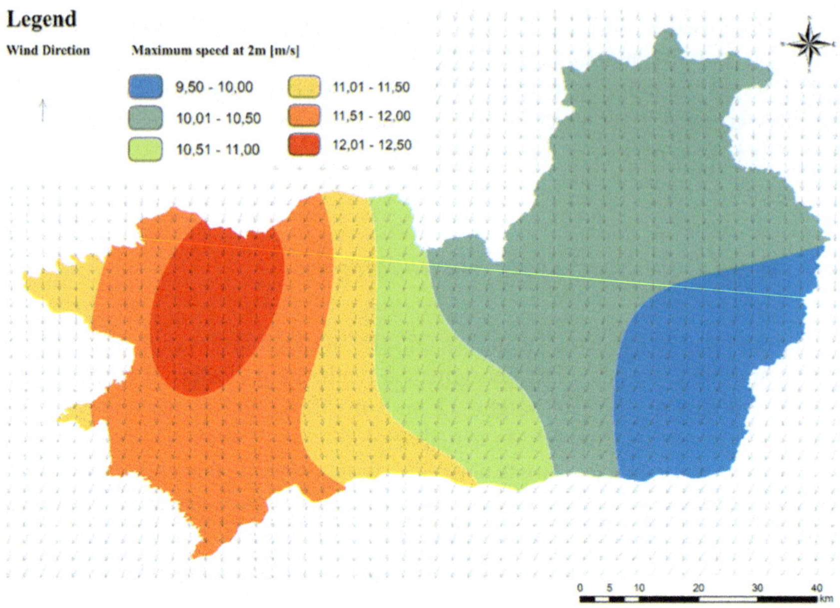

Fig. 4 Maximum speed (average 2001–2021)

3.1 *What is the Most Favorable Height for Producing Electricity?*

To calculate the wind intensity for the heights of 5, 10, and 20 m, the data was generated using Eq. (1), and the height for the Eq. (1) variability is shown in Table 1.

Models in Fig. 5 show the average annual wind speeds by height. The map shows the increase of wind speed with the rise in height; only the map of [2 m] height has wind direction, as it is the height of the ammeter. The predictive error of the model was analyzed, and it is presented in Appendix 1.

Having considered the wind speeds at the different heights, the wind power (Eq. 2) at each height is calculated. Bearing in mind that it is necessary to know the diameter of the wind turbine blades, the study adopted different diameters, namely, for a height of 5 m, a diameter (Ø) of 4 m, for H = 10 m ⇒ Ø = 12 m, and for H = 20 m ⇒ Ø = 30 m, guaranteeing clear spans of 3 m, 4 m, and 5 m, respectively, under the wind turbines.

Table 1 Calculation of wind speed at different heights

2001–2021	H_a [2 m]	H_a [5 m]	H_a [10 m]	H_a [20 m]
V_H [$m \cdot s^{-1}$] Min	0.96	1.15	1.32	1.52
V_H [$m \cdot s^{-1}$] Max	2.70	3.24	3.73	4.28

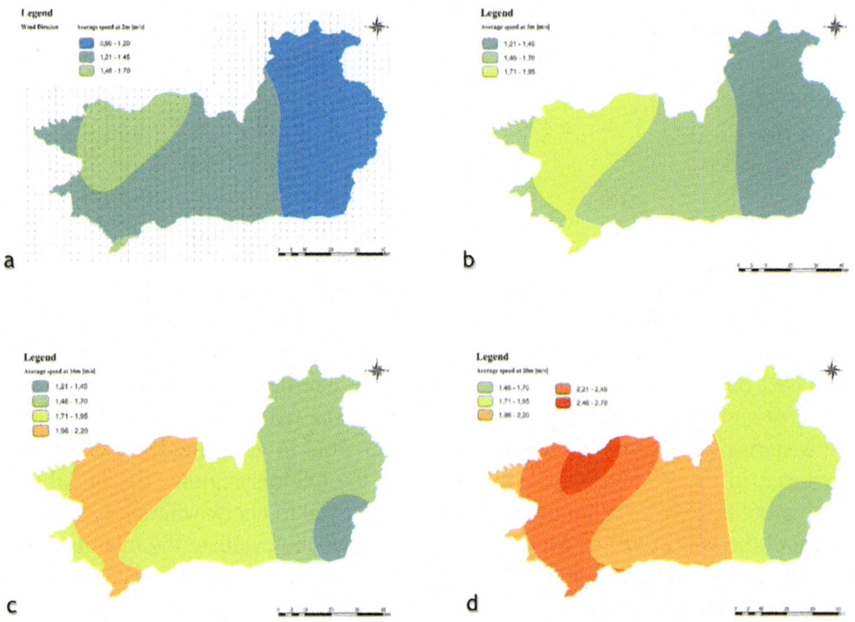

Fig. 5 Annual average speed (2001–2021): **a** H = 2 m + wind direction (WD); **b** H = 5 m; **c** H = 10 m; **d** H = 20 m

$$P_v = \frac{1}{2} \cdot \rho \cdot A \cdot v^3 \cdot P_v = \frac{1}{2} \cdot \rho \cdot \left(\frac{\pi \cdot r^2}{2} \right) \cdot v^3 \tag{2}$$

where: P_v is the calculated wind power [W]; ρ is the air density (1.225 [$kg \cdot m^{-3}$] at CNTP—T = 15 [°C], P = 1 [atm]); A is the area of the air section passing through the turbine's propellers; r is the radius of the turbine propellers; v is the wind speed.

The results of the wind power calculation for different heights, for the turbine diameters mentioned above, are shown in Table 2, with values of 0.53 [W] for the minimum wind speed $V_H = 0.96[m \cdot s^{-1}]$ (~3.5[$km \cdot h^{-1}$]).

In addition to demonstrating the annual energy potential, using Eq. (3), which calculates the energy potentially produced for the average daily wind values over one year (Table 3), it was observed that the worst production scenario occurs in the eastern portion of the CIMBB, as the average winds range from 0.96 to 1.2 [$m \cdot s^{-1}$] (~3.5 to 4.3 [$km \cdot h^{-1}$]).

Table 2 Calculation of wind power (Pv)

2001–2021	H_a [2 m]	H_a [5 m]	H_a [10 m]	H_a [20 m]
R[m]	0.5	2.2	6.0	15.0
P_V [W] Min	0.21	5.9	80.5	762.5
P_V [W] Max	4.73	131.3	1790.6	16,962.9

Table 3 Calculation of wind energy (Ev)

2001–2021	H_a [2 m]	H_a [5 m]	H_a [10 m]	H_a [20 m]
E_V [kWh/year] Min	0.7	20.7	282.1	2671.7
E_V [kWh/year] Max	16.6	459.9	6274.3	59,438.0

$$E_v = P_v \cdot CF \cdot 8.76 \quad (3)$$

where: E_v is the calculated potential energy generated by the wind [kWh/year]; P_v is the calculated wind power [W]; CF is the wind turbine capacity factor; 40% by default; 8.76 is the number of hours in 1 year divided by 1000 to obtain kWh

Based on the data computed in the tables, it can be seen that the height with the most significant generation is 20 m since, in the worst-case scenario, which is the eastern region of the CIMBB, it is still possible to generate 200 kWh/month. However, in the best-case scenario, which is the Oleiros region, it is possible to reach more than 500 kWh/month. That said, it is vital to check where it is desirable to install the wind turbine. Moreover, it is necessary to evaluate the manufacturers with the best technology, guaranteeing a high CF and thus offsetting the installation costs with higher production.

3.2 Distribution of Production Potential in an Average year (kWh)

The potential production of renewable energy is not only the evaluation of data from one energy source but also includes a hybrid system. The CIMBB has areas with water resources and a lot of sunshine. This complementary approach permeates a more comprehensive analysis of Beira Baixa's energy potential, particularly in mini and micro-generation scenarios.

To examine the potential for electricity production in the region, it is necessary to consider characteristics such as topography, the pattern, direction, and intensity of the winds, as well as solar incidence, the regularity of rainfall, and the characteristics of rivers and bodies of water.

Wind energy, which is the approach of this project, is not entirely constant, which causes moments of calm and, consequently, the absence of electricity production. Like solar panels, wind energy generation can store excess energy in batteries to be used during periods of calm.

The solution of a hybrid system with photovoltaic energy is welcome because the times when the winds intensify are at dusk and during the night due to the rising currents from the heated ground, reducing their intensity during the day, particularly in summer, when the incidence of the sun is greater due to the longer days. Currently, wind turbines are being manufactured with solar panels on their towers to make better use of this combination [16, 17].

Thus, generating cost savings and the possibility of selling surplus generation (profit). Such synergy is an important part of distributing the production potential of an average year and reduces dependence on a single source.

4 Final Remarks

In terms of the volume of electricity produced for each height covered in this study (2, 5, 10 and 20 [m]), it can be said that, without a doubt, the higher the altitude at which a wind turbine is placed, the greater the amount of energy it will produce, depending on the generator's power, which is linked to the length of the blades (a function of height), as well as the capacity factor (CF), which has increased as a result of technological advances and is currently close to 60%. Unfortunately, this means that equipment costs increase with increasing power and capacity factors.

That said, the integration of micro or mini production with a hybrid system for the development of decentralized energy management solutions and the creation of public incentive policies, namely specific credit lines for electricity production on these scales, offer the promise of an excellent future for Portugal in the energy area, guaranteeing the maintenance of sustainability in the Beira Baixa region.

The wind-based energy matrix stands out. Despite its moments of inconstancy, when combined with another form of renewable energy production, it facilitates the management of low productivity.

Studies show that decentralized energy generation in small production units is a promising approach to meeting local energy needs. However, how can micro and mini-generation technologies, such as photovoltaic panels and small wind turbines, be made more viable and desirable for farmers?

Using electricity in rural areas is not a convenience but a commodity. Producers, even small ones, have modernized and are buying more modern, less polluting machinery and equipment that require electricity instead of fossil fuels.

Acknowledgements The work was financed by FCT—Fundação para a Ciência e Tecnologia, I.P. to Research Centre for Natural Resources, Environment and Society (CERNAS-IPCB).

Appendix 1

Co-krigagem		
2 m	Predicted	0.0533705013512923x + 1.23615217992149
	Error	−0.946629498648708x + 1.23615217992150

(continued)

(continued)

	Standardized error		−2.66210114037576x + 3.47787206641089
	Optimize	Mean	−0.00882332035360645
		Root-Mean-Square	0.35881160467155800
		Mean standardized	−0.01904293046647680
		Root-Mean-Square Standardized	1.00463936880986000
		Average standard error	0.36054882165521300
5 m	Predicted		0.0534555658184429x + 1.48466794827222
	Error		−0.946544434181557x + 1.48466794827222
	Standardized error		−2.21619302148566x + 3.47768626541596
	Optimize	Mean	−0.01053709494220260
		Root-Mean-Square	0.43095765777011200
		Mean standardized	−0.01892540941433770
		Root-Mean-Square Standardized	1.00462835392518000
		Average standard error	0.43305406663533600
10 m	Predicted		0.0533804104759775x + 1.70556878790329
	Error		−0.946619589524023x + 1.70556878790329
	Standardized error		−1.92947436175227x + 3.47798059033243
	optimize	Mean	−0.01208990183075200
		Root-Mean-Square	0.49500070600596100
		Mean standardized	−0.01890435233873240
		Root-Mean-Square Standardized	1.00456374836261000
		Average standard error	0.49744580810175800
20 m	Predicted		0.0533843058107503x + 1.95922196492210
	Error		−0.946615694189249x + 1.95922196492210
	Standardized error		−1.67967483213759x + 3.47799931278687
	optimize	Mean	−0.01389270429096320
		Root-Mean-Square	0.56863885973478200
		Mean standardized	−0.01890936698537470
		Root-Mean-Square Standardized	1.00460068709759000
		Average standard error	0.57142249087597800

References

1. Silva A (2019) Optimização de Recursos Hídricos em Coordenação com a Produção Eólica: Decisão para Lances Conjuntos no Mercado do Dia Seguinte, Doctoral dissertation, Covilhã
2. Kariniotakis G et al (2006) Next generation forecasting tools for the optimal management of wind generation. In: Proceedings of international conference probabilistic methods applied to power systems, June, pp 1–6
3. Sánchez I (2006) Short-term prediction of wind energy production. Int J Forecast 22(1):43–56
4. Wu AJ, Guan X, Zhou X, Zhou Y (2011) Estimation and characteristic analysis of aggregated generation of geographically distributed wind farms. In: Proceedings of 2011 IEEE power engineering society general meeting, July, pp 1–6
5. Lew D, Milligan M, Jordan G, Piwko R (2011) The value of wind power forecasting. In: Proceedings of 91st American meteorological society annual meeting, 2nd conference weather, climate and new energy economy, Washington, DC, January
6. Potter C, Negnevitsky M, Jacka K (2004) Short-term wind forecasting application using an Adaptive Neural-Fuzzy Inference System (ANFIS). Artificial Intelligence in Science and Technology, pp 175–180
7. Roulston MS, Kaplan DT, Hardenberg J, Smith LA (2003) Using medium-range weather forecasts to improve the value of wind energy production. Renew Energy 28:585–602
8. Bremnes JB (2004) Probabilistic wind power forecasts using local quantile regression. Wind Energy 7(1):47–54
9. Juban J, Siebert N, Kariniotakis GN (2007) Probabilistic short-term wind power forecasting for the optimal management of wind generation. In: Proceedings of the IEEE powertech conference, Lausanne, Switzerland, July
10. Nielsen HA, Madsen H, Nielsen TS (2006) Using quantile regression to extend an existing wind power forecasting system with probabilistic forecasts. Wind Energy 9(1–2):95–108
11. Instituto Português do Mar e da Atmosfera, IPMA. https://www.ipma.pt/bin/file.data/climate-normal/cn_71-00_CASTELO_BRANCO.pdf. Accessed 08 Oct 2023
12. Instituto Português do Mar e da Atmosfera, IPMA. https://www.ipma.pt/bin/file.data/climate-normal/cn_81-10_CASTELO_BRANCO.pdf. Accessed 08 Oct 2023
13. Instituto Nacional de Estatística, ine.pt. https://www.ine.pt/. Accessed 08 Oct 2023
14. Sistema Nacional de Informação de Recursos Hídricos (SNIRH), snirh.pt. https://snirh.apambiente.pt/. Accessed 08 Oct 2023
15. Cerejo da Silva PDAJ (2021) Energia Eólica—O Vento, 2021. Avaliação do Recurso Eólico na Região da Beira Baixa com base nas Estações Meteorológicas do SNIRH 31
16. Pennsylvania State University—College of Earth and Mineral Sciences. https://www.e-education.psu.edu/emsc297/node/649. Accessed 08 Oct 2023
17. I'mNovation-hub. https://www.imnovation-hub.com. https://www.imnovation-hub.com/energy/new-generatiom-wind-turbines-photovoltaicpanels/?_adin=02021864894. Accessed 08 Oct 2023

Marginal Land Irrigated with Treated Wastewater. An Opportunity for Biofuel Production in Portugal?

Tânia Pires and Ana Maria Barreiros

Abstract The cultivation of industrial crops stands out as an alternative strategy for fossil fuels. However, it is important to ensure that the production of energy crops is not a threat to the production of food and feed, competing for land or water. This work was based on a bibliographical review of studies and projects that studied the potential for growing energy crops in marginal soils and the hypothesis of them being irrigated with wastewater. In fact, the use of land that no longer has the capacity for agriculture results in not only economic benefits but also environmental and economic ones. On the other hand, irrigation through the reuse of treated wastewater reduces pressure on water resources and adds the advantage of being able to deliver nutrients to energy crops. Irrigating energy plants with treated wastewater, from a proximity perspective with a defined area of 10 km in radius, allows some places to overcome the water needs of plants, leaving water for purposes other than energy generation.

Keywords Biofuels · Marginal soils · Treated wastewater

1 Introduction

Energy issues have been widely discussed, however the problem of energy security is far from resolved and tends to intensify with the imminent depletion of fossil energy sources; the increase in energy demand driven by population growth and purchasing

T. Pires (✉) · A. M. Barreiros
Instituto Superior de Engenharia de Lisboa—ISEL, Polytechnic University of Lisbon, Lisbon, Portugal
e-mail: tania.icpires@gmail.com

A. M. Barreiros
e-mail: ana.barreiros@isel.pt

A. M. Barreiros
UniRE—Unit for Innovation and Research in Engineering, Instituto Superior de Engenharia de Lisboa—ISEL, Polytechnic University of Lisbon, Lisbon, Portugal

P. S. Brito et al. (eds.), *Insights into Energy Trends*, Green Energy and Technology,
https://doi.org/10.1007/978-3-031-83811-8_8

power; security and lack of energy independence following global geopolitical instability and, among other concerns, the worsening of greenhouse gas (GHG) emissions and the respective consequences for climate change [1].

The energy sector, which includes transport, assumes important relevance, both economically and environmentally. This sector therefore faced with a complex challenge, that of guaranteeing access to energy in a sustainable way, without compromising the planet. In fact, transport, together with the production and transformation of energy, are characterized as the most representative sources of greenhouse gas emissions in Portugal [2]. Biofuels are considered carbon neutral with high potential for the decarbonization of the transport sector [3, 4]. That is one of the reasons why biofuels can be seen as an alternative source of energy.

In fact, international strategies and policies invoke the production and promotion of second-generation biofuels as a promising alternative to fossil fuels [1], representing an important contribution to meeting the Sustainable Development Goals (SDGs) of the 2030 Agenda [5].

Despite the positive role of biofuels in mitigating climate change and achieving the SDGs, they can simultaneously represent a threat, namely for food security [6]. In fact, first-generation biofuels use biomass from food crops, which is why it is necessary to promote production and consumption of next-generation biofuels, which do not compete with the production of food or animal feed, as second-generation biofuels. Second generation biofuels, also known as advanced biofuels, are produced from the biomass of energy crops, such as lignocellulosic species. Another issue associated with biofuels is the need of extensive areas for the growth of energy crops, the amount of water required for irrigation and possible use of fertilizers [7, 8].

This paradigm often leads to a trade-off mechanism, which imposes the choice and favor of one factor over another that will be penalized. The water-energy-food nexus approach argues that these resources must be viewed and managed in an integrated manner, highlighting the interconnection between water, energy and food, and aims to promote policies that consider the three sectors in an interdependent way [9]. The elements of the Water-Energy-Food Nexus are intrinsically related to each other, since the use of one of these resources requires and influences the others. In this context, the production of biofuels can be an alternative form of energy, without compromising food production and guaranteeing the preservation of the resources involved, it is essential to establish sustainable and efficient management of the means involved in their production, enhancing synergies and avoiding trade-offs [3, 10–12]. It is therefore important to carry out an analysis that allows efficient management of resources, particularly with regard to the location of cultivated land, which crops are most suitable for these locations, as well as the use of water, so that they do not represent threats between them, or for other resources.

In this context, much research has been done on the idea of using land without agricultural potential, generally known as marginal land, for the growth of energy crops aimed at producing biofuels or other forms of bioenergy without compromising food security, feed production or existing ecosystems [13, 14]. This approach also contributes to improving soil quality, and brings notable social, environmental and economic benefits [14]. In reality, the scientific community has produced a large number of studies on the production of energy crops on marginal soils, identifying and mapping marginal areas in European territory and selecting the crops most adapted to biophysical constraints, namely the MAGIC project—Marginal lands for Growing Industrial Crops (2017–2021) financed by the European Union [13–15].

MAGIC defined marginal lands as lands located in areas that once had agricultural potential, and that do not assume different functions such as urban uses, nature conservation, forestry or others. For that project, marginal soils were defined as "land with limitations that, as a whole, are severe for the sustained application of a given use and/or are sensitive to soil degradation as a result of inadequate human intervention, and/or already have lost some or all of their productive capacity as a result of inadequate human intervention and also include contaminated and potentially contaminated sites that constitute a potential risk to humans, water, ecosystems or other receptors". According to MAGIC, 29% of the European territory, which once had agricultural potential, is characterized as marginal land and, in mainland Portugal there is an area of 13,726 km^2 with biophysical restrictions, that is, 26% of the national territory, classified as agricultural between 1990 and 2012, it is considered marginal [16].

Regarding energy crops. the same project studied the most promising plant species for the production of biofuels, having initially addressed 67 industrial crops, based on databases and/or scientific projects. The consortium obtained the final list of 20 energy crops most suitable for the marginal soils under study through a multi-criteria analysis in which five criteria, with a pre-determined weight, were scored from 1 to 5 by the entire consortium. The criteria analyzed consisted of (a) experience with agricultural management of proposed industrial crops; (b) crop productivity for industrial applications (according to main uses); (c) expected performance of crops on marginal land; (d) industry needs and (e) market opportunities [17].

Finally, it is necessary to guarantee the efficient use of water in the irrigation of energy crops, not only due to competition with the irrigation of food, but also due to the increasingly accentuated water scarcity. In this context, the reuse of treated wastewater from WWTPs to irrigate energy crops presents itself as an opportunity with several associated benefits: reducing pressure on water resources, both in saving fresh water and in reducing discharges carried out in the receiving environment; reduction of fertilizers and also contributing to the recharge of aquifers with a lower load of contaminants. The reuse of wastewater playing an important role for the circular economy, in the sense that, in addition to the water itself, the nutrients present in these, such as nitrogen and phosphorus, are reused to fertilize the soil when reused in agricultural activity. Simultaneously, with the advantage of preventing the risk of environmental impacts such as pollution and eutrophication of the receiving environment where they are rejected.

Decree-Law No. 119/2019 [18] defines the legal framework for the production of water, originating from treated wastewater, for reuse through a "fit for purpose" approach. The mentioned approach says that the quality standards of water and reuse rules must be suitable for the purposes for which they are intended. In this case, the irrigation of energy crops falls into "Class D - Irrigation with restricted access (agricultural uses): seed production, including seeds for industrial use or energy production" and requires a "level of conventional secondary treatment and disinfection that includes UV radiation, ozonation, membrane processes, chlorination (only permissible for maintaining the disinfectant residual content) or other advanced oxidation processes" [19].

2 Methodology and Data Processing

The methodology adopted combined literature review and collection of information regarding the three components under study—marginal soils, energy crops and treated wastewater –, and data processing in ArcGIS software.

The information regarding marginal soils in Portugal was based on the work developed by the European project MAGIC, which identified, classified, and mapped marginal soils in Europe based on natural soil constraints, climate, and other topographic restrictions. The project made available information on marginal soils in a layer of geographic features, polygons that represent LAU 2, that is, parishes, classified according to the percentage of soil marginality. The layer referring to marginal soils was added to ArcGIS and, using the Filter tool, only soils located in Portugal (CNTR_CODE is PT) and with a marginal land area other than 0 (km^2 is not 0) were selected (Fig. 1).

The study of the most promising energy crops for marginal soils was also part of the MAGIC project, information that was also used in the present study. However, of the twenty crops selected by MAGIC [17], for this work, energy crops that could somehow compete with the production of food and feed, for example sorghum, were not considered; are identified as invasive in Portugal, namely acacias; or present

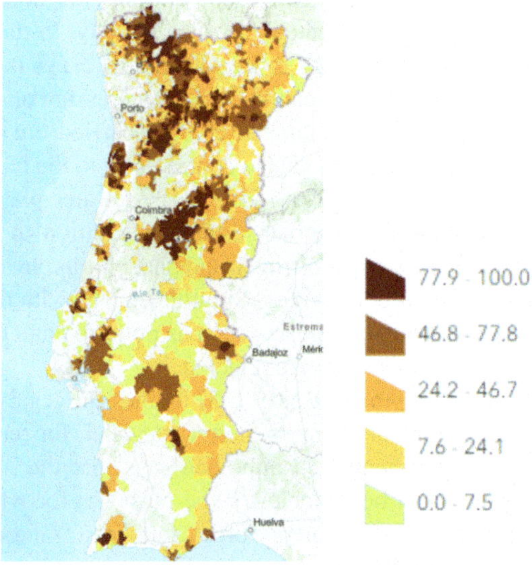

Fig. 1 Marginal soil layer edited in ArcGIS (% marginality by LAU 2)

legal plantation constraints, such as hemp [20]. Thus, the species under analysis were eight, whose characteristics are shown in Table 1.

Regarding the treated wastewater component, the necessary data on the location, type of treatment and capacity of the WWTP (p.e.) were obtained from the European Environment Agency website [23], based on mandatory reporting for population clusters with more of 2,000 inhabitants, within the scope of Directive 91/271/EEC of the European Council, of May 21, 1991 [24]. However, data were also collected from two WWTPs with population clusters, which reported even though it was not mandatory. The information, referring to 2020, was reported in 2022 by the EU-28 countries plus Iceland and Norway and awaits validation from the European Commission.

The European Environment Agency [23] provides interactive access to member states' reports and the feature layer hat has been added to the ArcGIS software. Once again, the Filter tool was used to select WWTPs located in the national territory with the potential to irrigate energy crops on marginal soils. In this phase, two scenarios were established: (i) first scenario in which only WWTPs with more advanced treatment than secondary were considered, as required in Decree-Law 119/2019 [18]; (ii) second scenario in which, in addition to WWTPs with more advanced than secondary treatment, WWTPs with only secondary treatment were additionally considered.

Thus, for Scenario 1, conditions were imposed that allowed the selection of WWTPs located in national territory with secondary treatment and with at least one advanced treatment (Fig. 2 Left). For Scenario 2, only the conditions relating to the location of the WWTP in the national territory and with secondary treatment were imposed (Fig. 2 Right).

In order to map the areas of marginal soil with the potential to be irrigated with effluents from the identified WWTP, a buffer was created in ArcGIS with a radius of

Table 1 Main characteristics of energy crops under study (adapted from [17, 21, 22])

Energy culture	Water requirement (mm/year or mm/growing season for annuals)	Productivity (t DM/ha/y)	Higher calorific value (MJ/kg)
Miscanthus	600–1000—adapted	40	17.5
Switchgrass	500–600—adapted 600–1000—very adapted	15	18.8
Reed canary grass	800–1000—adapted	4.7	17.3
Poplar	800–1000—adapted	7	19.6
Wild sugarcane	400–500—adapted 500–1000—very adapted	20	17
Willow	800–1000—adapted	20	17
Siberian Elm	400–500—adapted 500–1000—very adapted	7.5	18.2
Tall wheat grass	300–400—adapted 400–1000—very adapted	9.5	17.9

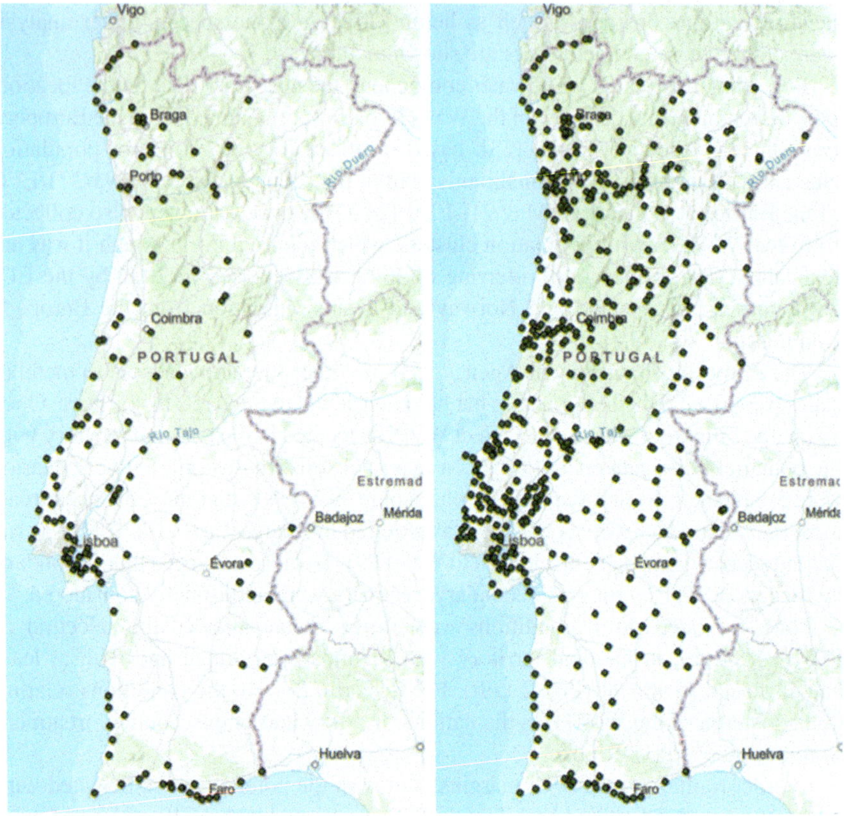

Fig. 2 Left—Scenario 1—ArcGIS map WWTP with more advanced treatment than secondary treatment; Right—Scenario 2—ArcGIS map WWTP with secondary treatment

10 km centered on each WWTP. The 10 km radius was defined based on the study on the potential for water reuse for irrigation in the European Union, which assumes that the transport of treated wastewater over distances greater than 10 km is not viable as it requires excessively high pumping efforts [25]. To this goal, through a proximity analysis, with the Create Buffer tool, the area with the potential to be irrigated with treated wastewater was obtained (Fig. 3).

The layers containing the buffers were superimposed on the marginal soil layers and, using the Intersect tool available in the Overlay Layers module, a new layer was obtained with the marginal soil areas that coincide with the area of 10 km radius from each WWTP. In order to give greater rigor to the area of marginal soils eligible for the cultivation of industrial species, layers of areas where it is not possible to cultivate were added, such as ZPE, SIC, RAMSAR, RNAC, Biosfera, bodies of water and built-up areas. These layers were intersected with the layer previously obtained, marginal soils within the perimeter of 10 km of the WWTP, and then these

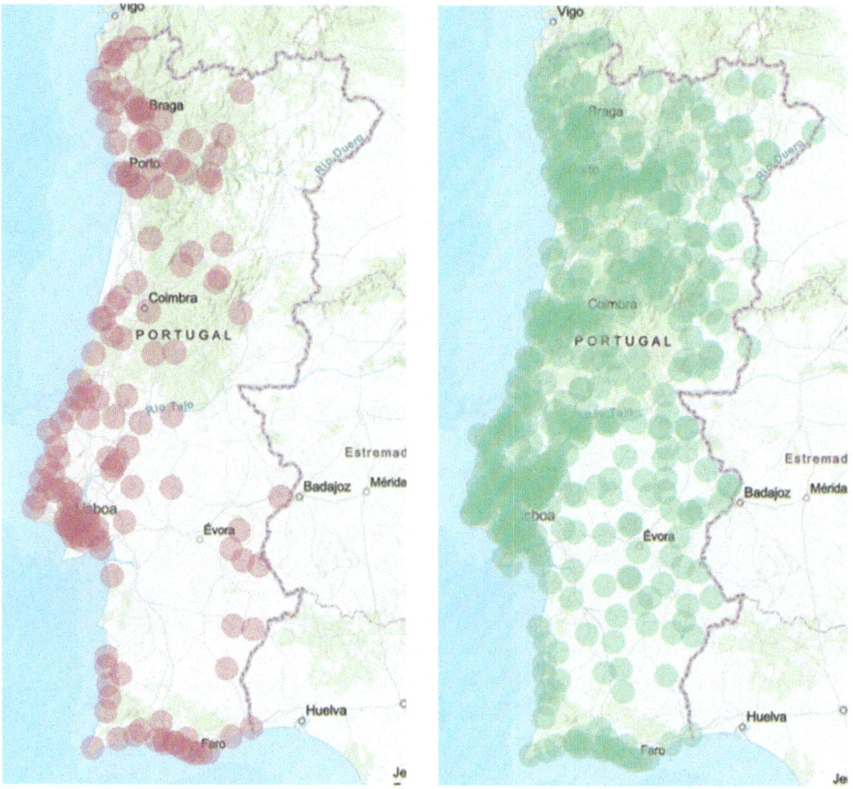

Fig. 3 ArcGIS layer—Buffer 10 km centered on each WWTP Left—Scenario 1; Right—Scenario 2

areas were eliminated using the Erase function of the Overlay Layers module. This procedure was carried out for both scenarios, obtaining the layers presented in Fig. 4.

Through the ArcGIS software, 2 layers were obtained with the areas eligible for the cultivation of energetic species, according to the type of treatment of the WWTP, with all the information relating to the soils in question and eligible WWTP, associated with the imported layers and edited in ArcGIS, they were exported to an excel table. In the excel file, for each of the scenarios under study, firstly, the amount of wastewater treated from each WWTP was estimated based on the capacity in p.e. and based on the capabilities defined in Regulatory Decree No. 23/95, of August 23 [26]. Then, knowing the capacity of treated wastewater available and the amount of water that one hectare of energy crops needs, the area that could be irrigated was calculated, imposing the logical condition that if the potentially irrigated area was greater than the area available, then the available area would be worth. Depending on this calculated value, knowing the productivity per hectare, the yield of each species was determined and therefore the energy value that the determined areas could potentially generate if planted with the species under study.

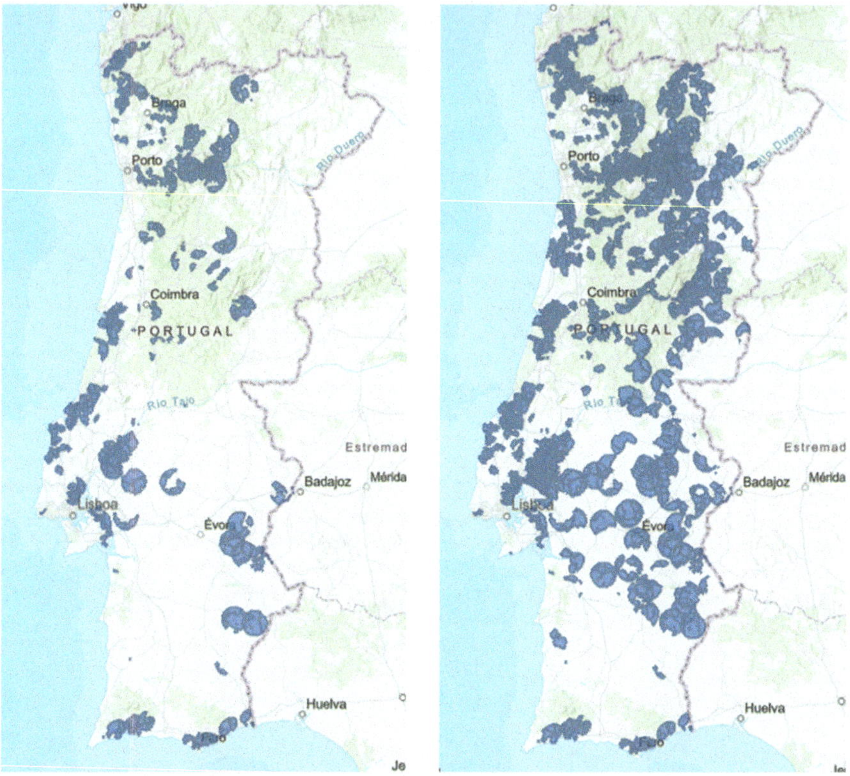

Fig. 4 ArcGIS layer—Marginal soils contained in the buffer 10 km of WWTP without uncultivable zones Left—Scenario 1; Right—Scenario 2

3 Results and Discussion

The assessment of the potential for reuse of treated wastewater on marginal soils for irrigation of energy crops was based on 3 aspects: water needs, productivity and energy generation.

Regarding the water needs of industrial species, the Magic project characterized the water needs of plants into classes, and for this study we analyzed the amounts of water that allow plants to fall into the "suitable" and "very suitable" classes. The greater the need for crop irrigation, the smaller the area that could be irrigated would be, however, the analysis carried out allowed us to verify that the water made available by some WWTP would respond not only to the intermediate needs of industrial species, but also to the optimal needs and there is still water left for other activities or services. Figure 5 represents area of eligible marginal soil with cultivation according to the water needs of the crops.

The second aspect studied is the productivity of industrial species. A single species could be grown on all the marginal soil that is suitable for irrigation using WWTP

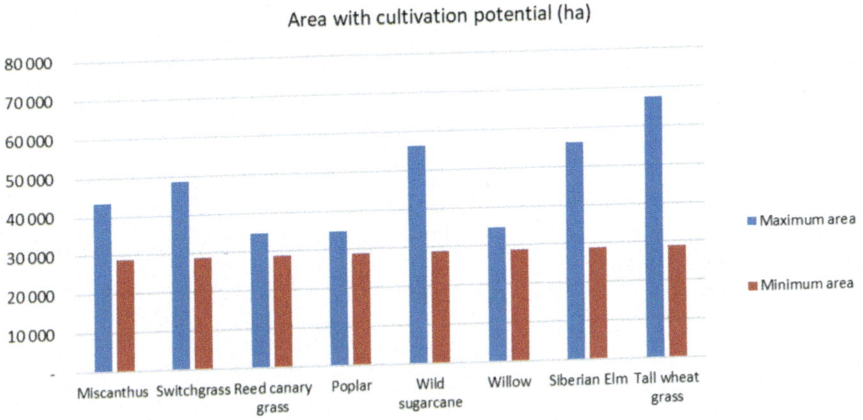

Fig. 5 Area of eligible marginal soil with cultivation potential, based on crop water needs

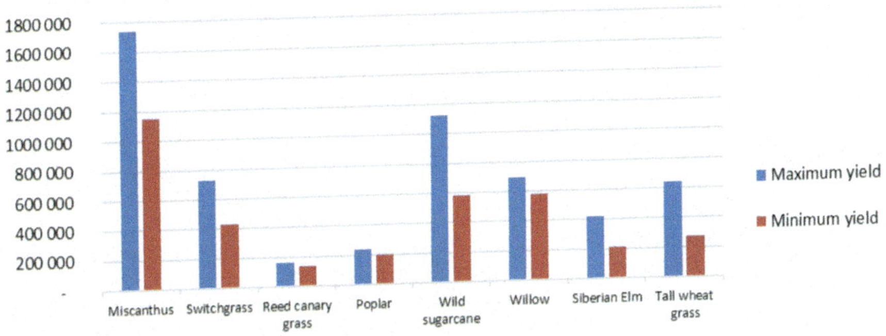

Fig. 6 Yield for the marginal area with production potencial (t DM/y)

water; Miscanthus would yield the highest yield while Reed canary grass would yield the lowest (Fig. 6).

Finally, the potential to generate energy in the area where each of the species could be cultivated was analyzed. Miscanthus and wild sugarcane are the crops that have the potential to generate more energy and reed canary grass and poplar are the plants that can generate less energy (Fig. 7).

4 Conclusions

Energy crops for biofuel production can be grown on marginal soils due to their low requirements. Their cultivation on marginal soils irrigated with treated wastewater is particularly advantageous because they grow more if they are irrigated. The crop with the highest yield and most potential for energy production among the eight

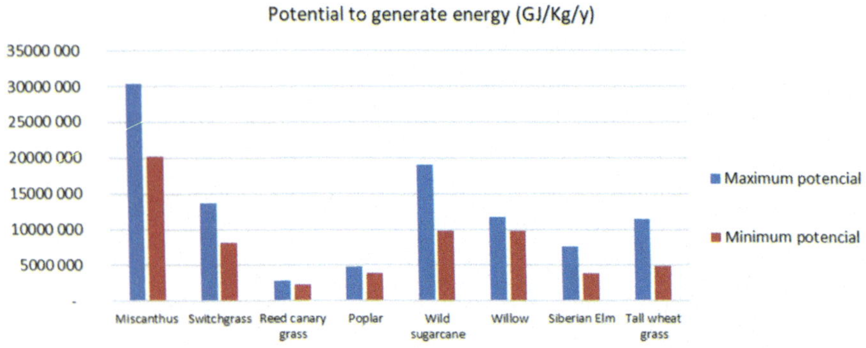

Fig. 7 Potential to generate energy (GJ/kg/y)

plants examined in this study was miscanthus. Despite WWTPs that produce more treated wastewater are near the cities and thus cover less marginal land area, this approach could be an opportunity for biofuels, especially if the issue of smaller WWTPs lacking treatment is overcome.

References

1. WBCSD: Biocombustíveis - Resumo Temático -Energia e Clima. https://bcsdportugal.org/wp-content/uploads/2013/11/publ-2008-Biocombustiveis.pdf. Last Accessed 04 Sept 2023
2. APA: Emissões de Gases com Efeito de Estufa. in Relatório do Estado do Ambiente. https://rea.apambiente.pt/content/emiss%C3%B5es-de-gases-com-efeito-de-estufa. Last Accessed 27 May 2023
3. Wang J et al (2023) Benefit analysis of multi-approach biomass energy utilization toward carbon neutrality. Innovation 4(3):100423
4. Ashokkumar V et al (2022) Recent advances in lignocellulosic biomass for biofuels and value-added bioproducts—A critical review. Bioresour Technol 344:126195
5. UN, «Agenda 2030 - ODS». Acedido: 27 de fevereiro de 2023
6. Nazari MT, Mazutti J, Basso LG, Colla LM, Brandli L (2021) Biofuels and their connections with the sustainable development goals: a bibliometric and systematic review. Environ Dev Sustain 23(8):11139–11156
7. Alesso SP, Tapias R, Alaejos J, Fernández M (2021) Biomass yield and economic, energy and carbon balances of Ulmus pumila L., Robinia pseudoacacia L. and Populus × euroamericana (Dode) Guinier short-rotation coppices on degraded lands under mediterranean climate. Forests 12(10):1337
8. Fernando AL, Costa J, Barbosa B, Monti A, Rettenmaier N (2018) Environmental impact assessment of perennial crops cultivation on marginal soils in the Mediterranean region. Biomass Bioenergy 111:174–186
9. Hoff H. The water, energy and food security Nexus, Stockholm Environment Institute (SEI), Stockholm. https://mediamanager.sei.org/documents/Publications/SEI-Paper-Hoff-UnderstandingTheNexus-2011.pdf. Last Accessed 23 June 2023
10. Lazaro LLB, Giatti LL, Bermann C, Giarolla A, Ometto J (2021) Policy and governance dynamics in the water-energy-food-land nexus of biofuels: proposing a qualitative analysis model. Renew Sustain Energy Rev 149:111384

11. FAO: The Water-Energy-Food Nexus A new approach in support of food security and sustainable agriculture. http://www.fao.org/3/a-bl496e.pdf. Last Accessed 18 March 2024
12. Foong SY et al (2022) The nexus between biofuels and pesticides in agroforestry: pathways toward United Nations sustainable development goals. Environ Res 214:113751
13. Abreu M et al (2020) Evaluation of the potential of biomass to energy in Portugal—Conclusions from the CONVERTE project. Energies 13(4):937
14. Khanna M et al (2021) Redefining marginal land for bioenergy crop production. GCB Bioenergy 13(10):1590–1609
15. Elbersen B et al (2019) Mapping Marginal land potentially available for industrial crops in Europe, March. https://doi.org/10.5281/ZENODO.2586947
16. Elbersen B et al (2018) D2.6 Methodological approaches to identify and map marginal land suitable for industrial crops in Europe. https://doi.org/10.5281/ZENODO.3539311
17. Alexopoulou E (2018) D1.3- List with the selected most promising industrial crops for marginal lands. https://doi.org/10.5281/ZENODO.3539150
18. Decreto-Lei n.o 119/2019 de 21 de agosto
19. APA: Guia para a reutilização de água - usos não potáveis: Agência Port. Ambiente (2019)
20. DGAV: Enquadramento Legal de Cânhamo (Cannabis sativa L.) e Produtos à Base de Cânhamo enquanto Alimentos para Animais (2023)
21. Alexopoulou E (2018) D1.5 – Handbook with fact sheets of the existing resource-efficient industrial crops. https://doi.org/10.5281/ZENODO.353916
22. Cossel M et al (2019) D4.1 Low-input agricultural practices for industrial crops on marginal land. https://doi.org/10.5281/ZENODO.3539368
23. Urban Waste Water Treatment map—European Environment Agency. https://www.eea.europa.eu/themes/water/european-waters/water-use-and-environmental-pressures/uwwtd/interactive-maps/urban-waste-water-treatment-maps-3
24. Directive nº 91/271/CEE, mai. 1991
25. European Commission (2017) Joint Research Centre: The potential of water reuse for agricultural irrigation in the EU: a hydro economic analysis. LU: Publications Office
26. Decreto Regulamentar nº 23/95

Characterization of Refuse-Derived Fuel Pellets and Feasibility Assessment of Their Application in CO_2 Gasification

Santa Margarida Santos, Catarina Nobre, Margarida Gonçalves, and Paulo Brito

Abstract In this study, RDF was pelletized, co-pelletized with lignocellulosic biomass (1:1), and then used as a feedstock in CO_2 gasification simulation tests using a TGA apparatus to verify its behavior and feasibility for syngas production. The RDF pellets had lower moisture content (5.42%) and better physical characteristics, such as durability (99.53%) and fines content (0.49%), suggesting that their physical integrity could be maintained during transportation and storage. Co-pelletizing improved the quality of the pellets, reducing the ash content and increasing the higher heating value. The gasification simulation tests showed that the RDF pellets with added biomass gasified more slowly and less linearly, whereas the RDF pellets converted faster and more uniformly.

Keywords Waste · Refuse-derived fuel · Pelletization · Gasification

1 Introduction

Industrialization and urbanization have resulted in an increase in energy demand accompanied by intensive use of resources, leading to a substantial production of solid waste. Directives imposed on European Union (EU) member countries have dictated the need to look for alternative sources of energy, such as waste [1]. The main advantages of using waste-derived fuels are a significant reduction in the volume of waste, the deviation of waste from landfills, and the possibility of energy recovery [2]. About 2.5 tons of steam, 0.5 MWh of electrical energy, and 21 kW of power can be produced with approximately one ton of Municipal Solid Waste (MSW) [3]. Refuse-derived fuel (RDF) is currently a material of interest in energy-intensive

S. M. Santos · M. Gonçalves
MEtRICs—Mechanical Engineering and Resource Sustainability Center, Department of Chemistry, NOVA School of Science and Technology, Caparica, Portugal

S. M. Santos · C. Nobre (✉) · M. Gonçalves · P. Brito
VALORIZA—Research Center for Endogenous Resource Valorization, Portalegre Polytechnic University, Portalegre, Portugal
e-mail: catarina.nobre@ipportalegre.pt

P. S. Brito et al. (eds.), *Insights into Energy Trends*, Green Energy and Technology,
https://doi.org/10.1007/978-3-031-83811-8_9

industries such as cement, paper, chemical production, or energy production [2]. RDF is produced through the processing of MSW in mechanical and biological treatment plants (MBT) and represents the combustible fraction of MSW, *i.e.* paper, cardboard, plastics, textiles, wood, leather, etc. Compared to other waste-to-energy (WtE) thermochemical technologies, the higher yield of gaseous fuel and the flexibility in varying the raw material make gasification an attractive choice for converting RDF [4]. Gasification is the partial oxidation at high temperatures of carbonaceous raw materials into syngas. Partial oxidation is carried out using different oxidizing agents (air, oxygen, steam, or their mixtures). Gasification is therefore classified according to the type of oxidizing agent used and the type of gasifiers (fixed bed, fluidized bed, and entrained flow) [3]. Gasification is approximately 69% technologically efficient, but this value changes significantly based on various factors, such as the raw material, reactor type, and operating parameters. The syngas quality is also significantly affected by these factors. The feedstock must be homogeneous and processed to specific sizes depending on the type of gasifier (1–100 mm), often requiring crushing and/or densification of the feedstock [5]. Some technical issues hinder the general use of RDF through the gasification technology, such as handling the high ash content (generally over 10 wt.%), moisture, and feedstock heterogeneity. The use of biomass together with RDF (e.g., co-gasification) has been applied as a measure to mitigate these technical issues [4]. Densification is another measure for improving the quality of RDF as a fuel. Pellets are often preferred in fuel applications owing to their improved physical properties (density, homogeneity, and shape) [6, 7].

According to García et al. [7], although the use of RDF as a fuel in thermal processes has been studied extensively in the literature, the opposite is true when it comes to pelletization or co-pelletization. According to the same author, waste-derived pellets can be successfully produced from RDF, with a durability of over 96.9% and a high heating value (HHV) above 10.7 MJ/kg [7]. In a study by Jewiarz et al. [8] it was found that the apparent density of RDF pellets could be up to five times higher than that of RDF fluff. Owing to the higher bulk density of RDF pellets, this fuel will be more beneficial in heat-treatment processes such as gasification [8].

In this study, RDF was pelletized, co-pelletized with lignocellulosic biomass (in a 1:1 proportion), and then used as a raw material in CO_2 gasification simulation tests using a thermogravimetric analyzer (TGA) apparatus to check the behavior and feasibility of this fuel for syngas production.

2 Experimental Setup

2.1 Raw Material

The raw material used in this work corresponded to RDF (supplied by Braval - Valorização e Tratamento de Resíduos Sólidos, S.A., Portugal) and end-of-life lignocellulosic biomass (provided by the company Casal & Carreira - Biomassa Lda),

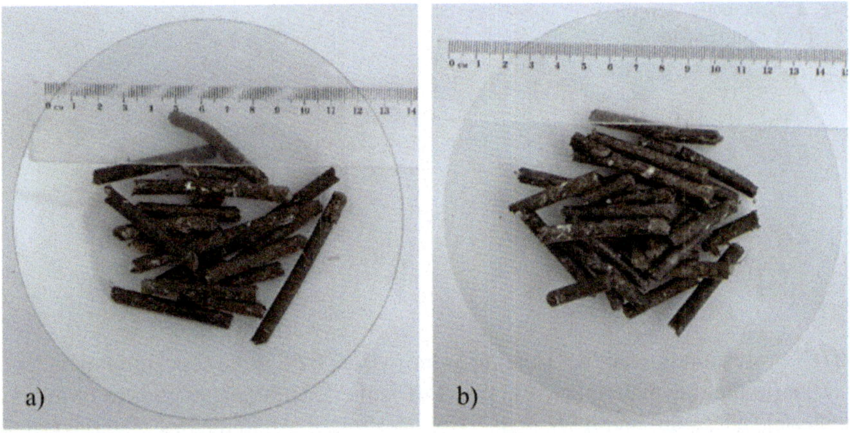

Fig. 1 Pellets obtained in this work and used for CO_2 gasification. **a)** RDF pellets; **b)** RDF:B pellets

consisting of a mixture of different forest residues from cleaning operations and demolition wood waste.

2.2 *Pelletization*

The composite RDF sample was crushed using a hammer mill of 1t/h capacity (Metalonicho, Italy) equipped with two screens of 10 and 2 mm. The end-of-life lignocellulosic biomass received was already crushed. The material was compacted in a pelletizer unit with 350 kg/h capacity (Andritz, Germany), equipped with a 6 mm die. Figure 1 shows the RDF pellets and RDF:B pellets (RDF and end-of-life lignocellulosic biomass mixed in proportions of 1:1).

2.3 *Characterization of the Produced Pellets*

The pellets were subjected to different physical–chemical characterization analysis, namely, proximate analysis, and high heating value (HHV) determination.

Moisture, volatile matter, and ash contents were determined according to standards ASTM E949-88 (fractions < 0.425 mm), ASTM E897-88 and ASTM E830-87, respectively. Fixed carbon was determined by difference in a dry basis (db). ***HHV*** was determined using Eq. (1) [9], as follows:

$$\boldsymbol{HHV}(\mathrm{MJ/kg,\,db}) = 0.3536\boldsymbol{FC} + 0.1559\boldsymbol{VM} - 0.0078\boldsymbol{ASH} \quad (1)$$

where:

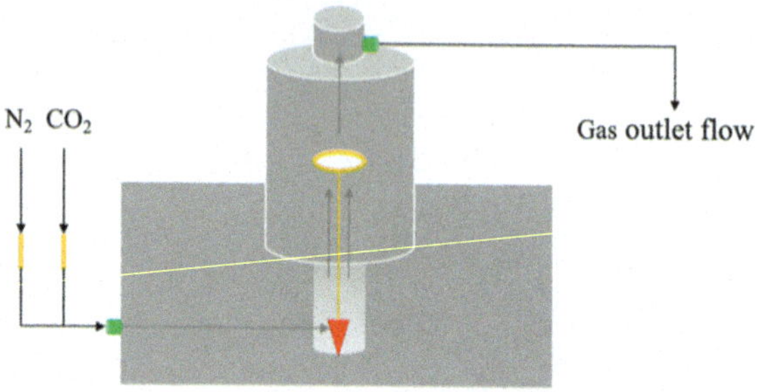

Fig. 2 Schematic view of the TGA setup. Adapted from [10]

HHV corresponds to Higher Heating Value (MJ/kg), FC is the concentration of fixed carbon in the raw material (wt.%, db), ***VM*** is the volatile matter content of the raw material (wt.%, db) and ***ASH*** is the concentration of ash in the raw material (wt.%, db).

Bulk density was obtained following EN 15103:2009 (EN ISO 17828:2015). Mechanical durability and amount of fines (<3.15 mm) were determined using EN ISO 17831-1:2016.

2.4 CO_2 Gasification Using RDF and RDF:B Pellets

The experimental setup used for gasification studies is schematically shown in Fig. 2. The gasification experiments were carried out using CO_2 at 950 °C in isothermal conditions with RDF pellets and RDF:B pellets using a NETZCSH ST490 F3 thermogravimetric analyzer (TGA). The pellets were heated at a constant heating rate of 20 °C/min in pure N_2, from room temperature to 950 °C. The temperature ramp was followed by a 10 min isothermal phase. CO_2 (15%) was then injected while the total gas flow and temperature were kept constant. The CO_2 gasification continued at a constant temperature until no further mass loss was observed.

3 Results and Discussion

3.1 Pellet Characterization

The characterization of RDF and RDF:B pellets is represented in Table 1.

The addition of biomass to RDF led to a slight increase in the moisture content of the pellets, although both pellets had a low moisture content within the levels recommended by ENplus® for commercial wood pellets ($\leq$ 10 wt.%) [11]. The opposite effect was observed for the ash content with the addition of biomass to the RDF. The ash content is one of the most relevant parameters for evaluating

Table 1 Summary of the physical–chemical characteristics of the produced pellets

Parameters	Units	RDF pellets	RDF:B pellets
Moisture	wt.%, ar	5.42 ± 0.16	7.64 ± 0.18
Volatile matter	wt.%, db	79.94 ± 0.81	74.20 ± 1.09
Fixed carbon		0.94 ± 0.42	10.30 ± 0.95
Ash		19.13 ± 0.63	15.50 ± 0.18
HHV	MJ/kg, db	21.40	23.02
Bulk density	kg/m^3	551.10	567.73
Durability	wt.%,ar	99.53	98.56
Fines amount	wt.%, ar	0.49	1.37

the potential use of raw materials for energy production, given the problems of ash agglomeration inside the reactor, deposit formation, corrosion, fouling, and emission problems that are characteristic of raw materials with a high ash content [12]. RDF:B pellets had a lower ash content compared to RDF pellets (19% reduction). According to Pohlmann et al. [13] the high ash content is associated with the heterogeneity of the raw materials [13]. Biomass typically has a lower ash content than MSW or RDF, so co-pelletization ends up being beneficial as it moderates the issue of RDF's high ash content [14]. In addition, according to Pio et al. [15], besides the addition of biomass to RDF as a strategy to reduce the ash content of RDF (making it suitable for applications such as gasification), the addition of biomass can lead to benefits in terms of syngas quality and tar formation, contributing to solving recognized process problems associated with the gasification of RDF. Within these problems, the fact that RDF is rich in plastics, deeply affects feeding systems whilst contributing to an increased formation of contaminants [15]. The ash content found in both pellet types produced in this work, would allow them to be used as raw materials in fluidized bed gasification reactors, where according to Arena [16], the recommended ash content in the raw material can be up to 25 wt.% [16]. As seen in Table 1, the calculated HHV of RDF and RDF:B pellets is 21.40 and 23.02 MJ/kg, respectively. Thus, there was an increase in this parameter with the addition of biomass, probably because of an increase in the fixed carbon content. This result is in line with that reported by Garcia et al. [7], where it was found that a higher percentage of biomass in RDF pellets resulted in a higher HHV [7]. Nevertheless, more recent studies published by other authors, such as Chotikhun et al. [14] and Laosena et al. [17], disagree. These two authors reported that a higher concentration of RDF in pellets and a lower concentration of biomass resulted in a higher HHV. This discrepancy in the results can be explained by the origin and heterogeneity of the RDF and the type of lignocellulosic biomass used in the blend, which may have a higher or lower HHV. Laosena et al. [17] reported that pellet samples with the highest plastic content had the highest heating values [7, 17]. With regard to durability and the amount of fines, the results showed that, overall, both pellets were of high quality, with the RDF pellets showing the best results. RDF pellets had higher durability (99.5%) and lower fines content (0.49%) than RDF:B (98.5% and 1.37%, respectively). In a similar study,

Chotikhun et al. [14] found that a greater presence of RDF favored pellet durability [14]. According to Jewiarz et al. [8], biomass is more brittle than RDF, which explains the higher fines content in RDF:B pellets [8]. This suggests that RDF:B pellets during transportation and storage may break more easily, potentially compromising their integrity compared to RDF pellets. Overall, the addition of biomass improves the quality of the pellets, reducing the ash content and increasing HHV. RDF pellets have a lower moisture content and better physical characteristics, such as durability and fines content. These results are a strong indicator that this feedstock has potential to be used in the gasification process.

3.2 CO_2 Gasification Using RDF Pellets and RDF:B Pellets

Figure 3 presents the results obtained in the CO_2 gasification tests using RDF and RDF:B pellets.

For the two pellet samples, there were three significant mass loss peaks. The RDF:B pellets exhibited peak mass losses at 16 min (350 °C) and 22 min (around 470 °C). According to Robinson et al. [18], the first mass loss peak corresponds to the decomposition of the biomass present, namely the molecular structures such as hemicellulose and cellulose. The latter decomposes completely at 355 °C. The RDF pellets showed a more significant peak at 22 min (around 470 °C), matching the second peak in the mass loss of the RDF:B pellets. This decomposition may be due to the decomposition of the lignin present in the biomass and the plastic material present in the RDF, as the plastics begin to decompose at 440 °C [18, 19]. According to a study by Chhabra et al. [19], 470 °C, the temperature at which there was a significant loss of mass in the two pellet samples, is the temperature at which most of the components of low-density polyethylene (LDPE) are expected to volatilize to form gas [19].

Considering the timeline (in minutes) required for complete gasification, the RDF pellets took 5 min from the start of gasification (the stage at which CO_2 was injected) until the mass loss of the material stabilized, whereas RDF:B pellets took 8 min. In terms of the behavior during gasification with CO_2, the RDF pellets with added biomass exhibited less linear behavior and a longer conversion time. On the other hand, RDF pellets seem to have a faster and more uniform conversion.

4 Conclusions

This paper presents an experimental investigation of the production and characterization of RDF pellets, RDF and biomass blended pellets and the analyses of their feasibility as feedstock for CO_2 gasification. The RDF pellets presented a lower moisture content and better physical characteristics, such as durability and fines content, whereas the addition of biomass improved the quality of the pellets, reducing the ash

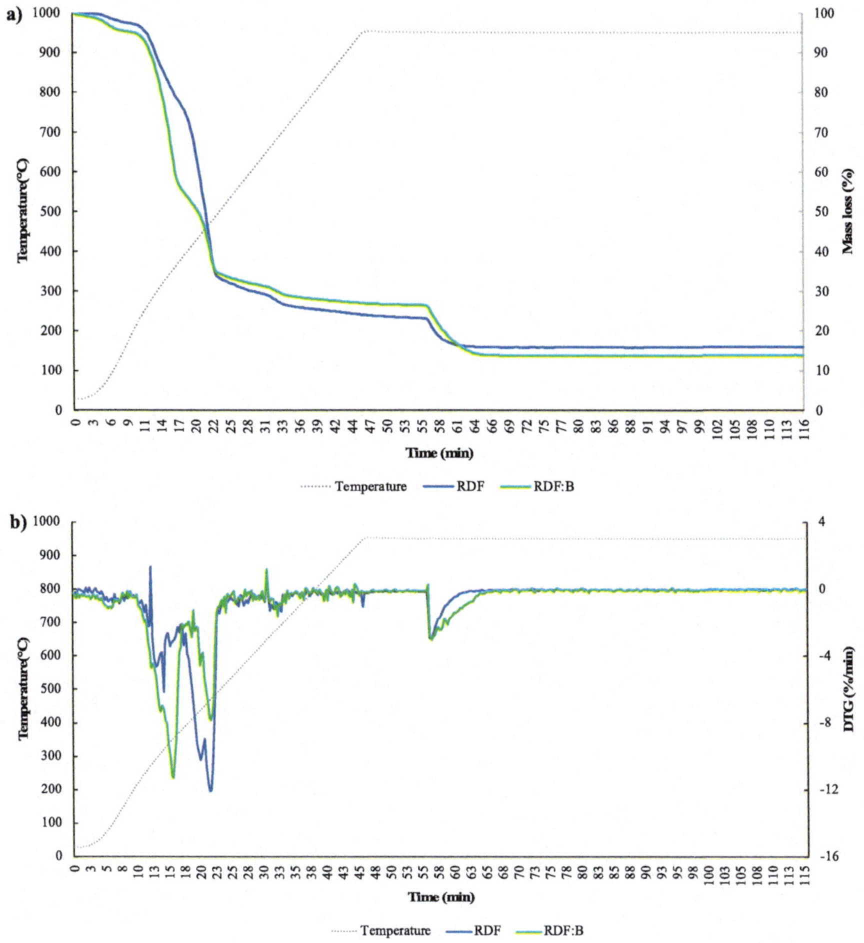

Fig. 3 TGA (**a**) and DTG (**b**) curves for RDF and RDF:B pellets showing isothermal CO_2 gasification at 950 °C

content and increasing their calorific value. The CO_2 gasification of the two-pellet samples showed that the RDF:B pellets seemed to behave less linearly and took longer to gasify, while the RDF pellets seemed to convert faster and more uniformly. Characterization of the pellets showed that through pelletization and co-pelletization, it is possible to process a waste stream, RDF, turning it into a raw material with suitable characteristics for gasification.

Acknowledgements This research was funded by Fundação para a Ciência e a Tecnologia, I.P. (Portuguese Foundation for Science and Technology), grant 2022.09990.BD, by project UIDB/05064/2020 (VALORIZA – Research Centre for Endogenous Resource Valorization), FCT/MCTES UIDB/04077/2020–2023 and UIDP/04077/2020–2023 (MEtRICs—Mechanical Engineering and

Resource Sustainability Center). This work was also supported by project WASTE2H2 (European Union's Horizon 2020 research and innovation programme) under grant agreement No 952593.

References

1. Zawiślak K et al (2020) The use of lignocellulosic waste in the production of pellets for energy purposes. Renew Energy 145. https://doi.org/10.1016/j.renene.2019.06.051
2. Gallardo A, Carlos M, Bovea MD, Colomer FJ, Albarrán F (2014) Analysis of refuse-derived fuel from the municipal solid waste reject fraction and its compliance with quality standards. J Clean Prod 83. https://doi.org/10.1016/j.jclepro.2014.07.085
3. Amin N et al (2024) Municipal solid waste air gasification using waste marble powder as a catalyst for syngas production. J Energy Inst 113. https://doi.org/10.1016/j.joei.2023.101496
4. Fazil A, Kumar S, Mahajani SM (2023) Gasification and Co-gasification of paper-rich, high-ash refuse-derived fuel in downdraft gasifier. Energy 263. https://doi.org/10.1016/j.energy.2022.125659
5. Makwana J, Dhass AD, Ramana PV, Sapariya D, Patel D (2023) An analysis of waste/biomass gasification producing hydrogen-rich syngas: a review. Int J Thermofluids 20. https://doi.org/10.1016/j.ijft.2023.100492
6. Wang T et al (2018) Evaluation of the potential of pelletized biomass from different municipal solid wastes for use as solid fuel. Waste Manag 74. https://doi.org/10.1016/j.wasman.2017.11.043
7. García R, González-Vázquez MP, Rubiera F, Pevida C, Gil MV (2021) Co-pelletization of pine sawdust and refused derived fuel (RDF) to high-quality waste-derived pellets. J Clean Prod 328. https://doi.org/10.1016/j.jclepro.2021.129635
8. Jewiarz M, Mudryk K, Wróbel M, Fraczek J, Dziedzic K (2020) Parameters affecting RDF-based pellet quality. Energies (Basel) 13(4). https://doi.org/10.3390/en13040910
9. Parikh J, Channiwala SA, Ghosal GK (2005) A correlation for calculating HHV from proximate analysis of solid fuels. Fuel 84(5):487–494. https://doi.org/10.1016/j.fuel.2004.10.010
10. Ge Y, Ding S, Kong X, Kantarelis E, Engvall K, Pettersson JBC (2023) Online monitoring of alkali release during co-pyrolysis/gasification of forest and agricultural waste: element migration and synergistic effects. Biomass Bioenergy 172. https://doi.org/10.1016/j.biombioe.2023.106745
11. European Pellet Council (2015) Quality certification scheme for wood pellets
12. Gomez-Martin A, Chacartegui R, Ramirez-Rico J, Martinez-Fernandez J (2018) Performance improvement in olive stone's combustion from a previous carbonization transformation. Fuel 228(February 2017):254–262. https://doi.org/10.1016/j.fuel.2018.04.127
13. Pohlmann JG, Osório E, Vilela ACF, Diez MA, Borrego AG (2014) Integrating physicochemical information to follow the transformations of biomass upon torrefaction and low-temperature carbonization. Fuel 131:17–27. https://doi.org/10.1016/j.fuel.2014.04.067
14. Chotikhun A et al (2023) Elemental compositions of wood plastic pellets made from sawdust and refuse-derived fuel (RDF) waste. Appl Sci 13(20):11162. https://doi.org/10.3390/app132011162
15. Pio DT, Tarelho LAC, Tavares AMA, Matos MAA, Silva V (2019) Co-gasification of refused derived fuel and biomass in a pilot-scale bubbling fluidized bed reactor. Energy Convers Manag 206(October):2020. https://doi.org/10.1016/j.enconman.2020.112476
16. Arena U (2012) Process and technological aspects of municipal solid waste gasification. A review. Waste Manag 32(4):625–639. https://doi.org/10.1016/j.wasman.2011.09.025
17. Laosena R et al (2022) Characterization of mixed pellets made from rubberwood (Hevea brasiliensis) and Refuse-Derived Fuel (RDF) waste as pellet fuel. Materials 15(9). https://doi.org/10.3390/ma15093093

18. Robinson T, Bronson B, Gogolek P, Mehrani P (2016) Sample preparation for thermo-gravimetric determination and thermo-gravimetric characterization of refuse derived fuel. Waste Manag 48:265–274. https://doi.org/10.1016/j.wasman.2015.11.018
19. Chhabra V, Bhattacharya S, Shastri Y (2019) Pyrolysis of mixed municipal solid waste: Characterisation, interaction effect and kinetic modelling using the thermogravimetric approach. Waste Manag 90:152–167. https://doi.org/10.1016/j.wasman.2019.03.048

Hydrotreated Vegetable Oil (HVO): A Review on Production, Properties, Environmental Impact, and Future Perspectives

Fernanda Macedo de Araujo Azeredo, Cristina Moll Hüther, Julia Barbosa Werlang, and Roberta Panizio

Abstract The increasing prominence of clean and sustainable energy sources drives the search for technologies and fuels that meet environmental demands and promote sustainable development. Renewable raw materials, such as biofuels, emerge as viable alternatives to reduce dependence on fossil fuels and mitigate greenhouse gas emissions. Among the most used biofuels, hydrotreated vegetable oil (HVO), also known as green diesel, stands out, with significant projections for increased production by 2030. The replacement of oil by renewable sources, encouraged by various nations, seeks to minimize environmental impacts compared to conventional fuels. In this context, HVO emerges as a promising option, offering environmental advantages and being produced from vegetable oils. This review highlights the importance of HVO in biofuel production, addressing its production, properties, and environmental impact, as well as its future prospects. Despite the advantages, challenges in the use of this energy are identified. Continuous investments in research, development, and infrastructure are crucial to maximize the potential of sustainable energy solutions. For this purpose, consultations were carried out on articles available in databases. This search was conducted using descriptive terms in both Portuguese and English, such as: "HVO", "green diesel", "hydrotreated vegetable oil", "biofuels", "alternative fuel" and "future prospects of biofuels", selected from the vocabulary found in the articles.

Keywords Biofuel production · Vegetable oils · Renewable energy

F. Macedo de Araujo Azeredo (✉) · R. Panizio
Polytechnic Institute of Portalegre, Portalegre, Portugal
e-mail: azeredofernanda@id.uff.br

C. Moll Hüther
Department of Agricultural and Environmental Engineering, Fluminense Federal University, Rio de Janeiro, Brazil

J. Barbosa Werlang
Federal Rural University of Rio de Janeiro, Rio de Janeiro, Brazil

P. S. Brito et al. (eds.), *Insights into Energy Trends*, Green Energy and Technology,
https://doi.org/10.1007/978-3-031-83811-8_10

1 Introduction

The use of clean and sustainable energy sources is gaining prominence and importance in the field of energy generation. Every day, efforts are made to discover new technologies and fuels that satisfy the requirements for reducing pollutant emissions, carbon dioxide emissions, and encourage sustainable development [1]. Currently, fossil fuels represent the main source of energy and chemicals, with about 80% of global energy consumption derived from their combustion, with the transportation sector being one of the largest contributors to this statistic [2]

In this scenario, renewable raw materials have proven relevant to decrease the dependence on the oil industry, reduce greenhouse gas emissions, and create new jobs in the agricultural sector [3]. Biofuels are derived from renewable organic matter and represent a viable alternative to replace, totally or partially, fuels derived from petroleum and natural gas in various energy generation applications [4]. Among the most used renewable energies derived from biomass are biodiesel and ethanol, whose production has gained significant projection [3]. It is estimated that by 2030, the production of these biofuels should reach 6.5 million barrels per day (Mb/d), a notable increase compared to the 1.8 Mb/d recorded in 2010 [2].

The European Community, the United States, and several other countries have been encouraging the replacement of petroleum with fuels from renewable sources. As expected, the environmental impacts resulting from these fuels tend to be smaller compared to the damage caused by fossil fuels, such as conventional diesel [5]. Biodiesel, the traditional renewable substitute for petroleum diesel, is produced from the conversion of vegetable oils and animal fats through the process of transesterification of triglycerides with methanol (see Fig. 1) [6]. However, due to its physicochemical characteristics and the existing transportation infrastructure, which is widely adapted for liquid hydrocarbons derived from petroleum, biodiesel cannot completely replace conventional diesel [7].

An increasingly considerable alternative for direct and more facilitated replacement is the use of fuels known as "drop-in," which can be employed in conventional vehicles without the need for mechanical alterations, and without demanding modifications in the existing infrastructure for storage and transportation [9]. In this sense,

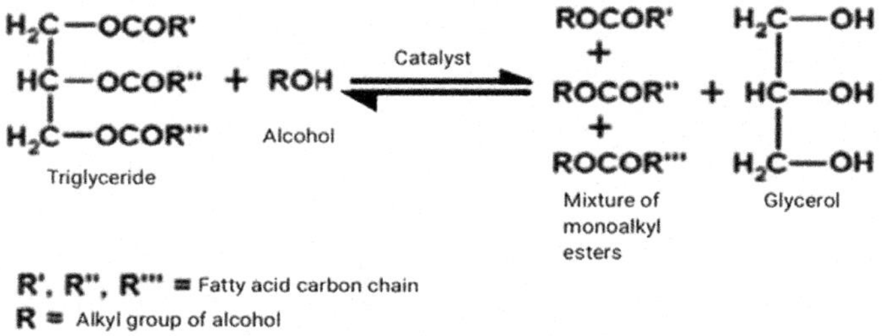

Fig. 1 Transesterification model adapted from Racaczeski et al. [8]

green diesel, also known as HVO (Hydrotreated Vegetable Oil), emerges as a particularly promising option in replacement of conventional diesel oil. HVO is a paraffinic biofuel produced from a variety of vegetable oils. It can be used in conventional diesel engines, either pure or mixed with fossil diesel [10]. These characteristics make it an attractive and effective option to mitigate the environmental impacts associated with the use of petroleum-derived fuels [6].

Given this panorama, the objective of this review is to present and highlight the growing importance of hydrotreated vegetable oil (HVO) as a promising alternative in the context of biofuel production and points out how production is done, its properties, the environmental impact, and future prospects of this product. The search for cleaner and more sustainable energy sources has driven research and development of technologies like HVO. Investing in studies and research on green diesel and other similar technologies is essential to face global challenges related to energy and the environment, ensuring a more promising future for the coming generations.

2 HVO Production

The production of HVO involves a series of complex steps that range from the careful selection of raw materials to the final fuel obtained through the hydrotreatment process (see Fig. 2). Initially, vegetable oils or animal fats are chosen, such as those from soy, sunflower, peanut, among others, as well as residual oils [11, 12]. These raw materials undergo a rigorous purification process to remove impurities and contaminants, using techniques such as neutralization, degumming, and bleaching [10].

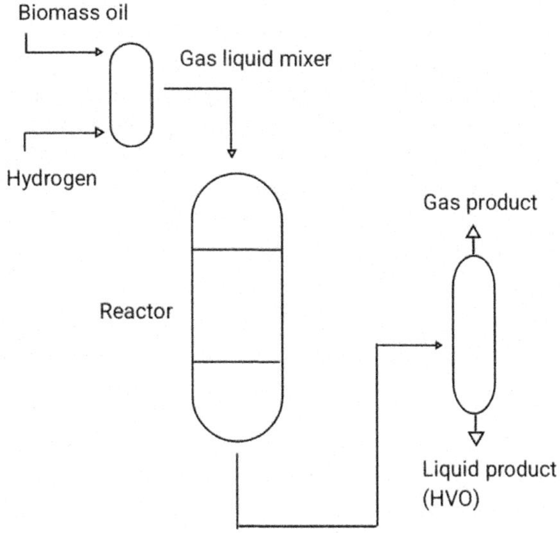

Fig. 2 Vegetable oil hydrotreatment process for HVO production. Image adapted from Sonthalia and Kumar [12]

Subsequently, the purified oil or fat is subjected to the catalytic hydrotreatment process. The hydrogen required for this step can be obtained from various sources, including oil refineries, natural gas reforming, or even water electrolysis [6]. However, the use of natural gas as a source of hydrogen raises questions about the sustainability of HVO as a renewable fuel [13]. During hydrotreatment, the biomass and hydrogen mixture, in the presence of a bifunctional catalyst, is carried out at temperatures of 300–370 °C and pressures of 2–10 MPa, the most important parameters in the reaction [14]. It is at this moment, due to the excess hydrogen present in the mixture, that promotes the breaking of double bonds between carbons and the formation of single bonds [12, 14].

Following the deoxygenation stage, oxygen molecules are removed through decarboxylation, decarbonylation, and hydrodeoxygenation reactions. These reactions convert triglycerides into paraffinic chains, eliminating oxygen in the form of propane (C_3H_8), carbon dioxide (CO_2), and water (H_2O). After this phase, the products are separated from the reactor, resulting in a hydrotreated vegetable oil with characteristics similar to conventional diesel [11].

The typical outcome is a liquid blend of straight-chain hydrocarbons that boils within the diesel range. By-products like CO, CO_2, and H_2O signify a reduction in atomic (mass) effectiveness. This green diesel, once produced, can be stored and transported for utilization in standard diesel engines, either alone or alongside traditional diesel fuel [15].

3 HVO Properties

In this section, the properties of hydrotreated vegetable oil (HVO) that offer significant advantages over conventional fuels will be presented, as well as some disadvantages to consider. A detailed understanding of these characteristics is crucial for a comprehensive assessment of HVO's potential in the current energy context. By exploring its properties, we can better understand how this fuel alternative can contribute to mitigate environmental impacts. Additionally, by identifying potential disadvantages, we can adopt more informed approaches for its implementation and sustainable development. This critical analysis of HVO's properties is essential to guide public policies, research and development investments, and strategic decisions in the energy sector [16].

3.1 Advantages

Hydrotreated vegetable oil primarily consists of paraffin and free aromatic compounds, oxygen, and sulfur, resulting in a notable reduction in greenhouse gas (GHG) emissions. Moreover, HVO has a cetane rating and calorific value that demonstrate superior performance to biodiesel in terms of energy efficiency. Its flash point

Table 1 Physicochemical properties of HVO, biodiesel and conventional diesel fuel. Table adapted from Simacek et al. [19]

	Density [kg.m^{-3}]	Viscosity at 40 °C [mm^2/s]	Distillation[°C]	Nitrogen [mg.kg^{-1}]	Sulphur [mg/kg]	Total aromatics [% mass]
Diesel	833.5	2.66	351	34	8	25.0
Biodiesel	882.6	4.55	354	8	4	<0.1
HVO	778.6	2.97	296	2	<2	0.2

varies within a specific range, between 68 and 120 °C, which facilitates its handling and storage [17, 18]. The operational conditions of this advanced biofuel can be adjusted and optimized to accommodate different raw materials, as well as the use of different catalysts and seasonal variations in the cloud point specifications of diesel [1]. Other physicochemical properties of all evaluated pure fuels are presented in Table 1 [19].

The quality of HVO is not directly linked to the origin of the raw material, and its low-temperature flow characteristics can be controlled by adjusting the severity of the reactor [1, 18]. The green diesel production process generates a gas effluent rich in propane, which can be easily separated from the fuel but has a lower value compared to glycerol. Thus, the propane-rich gas can be used for cogeneration of energy or as a reagent in the process, reducing the use of non-renewable fuels by providing an internal source of hydrogen or heat/electricity [7]. Moreover, hydrotreated vegetable oil has greater operational flexibility due to the possibility of adapting the process to control the length of the hydrocarbon chains—through hydrocracking—and to increase the cold flow properties—with isomerization. These steps are important for obtaining higher quality diesel and green jet fuel [20].

Finally, it is worth noting that the HVO production process can be conducted in existing refineries, without the need for new plants designed for this purpose or adapted [21]. This not only reduces the costs and time required for its large-scale production but also represents a more efficient solution in terms of infrastructure. Due to these characteristics, HVO is called a "drop-in" fuel and can be used in pure form in any proportion with conventional diesel, without the need for mechanical modification, this offered flexibility increases its attractiveness as a viable transition solution of all the evaluated pure fuels [19, 21].

3.2 *Disadvantages*

The main disadvantage lies in its high production cost, which is largely due to the raw material used, as large-scale production of these vegetable oils often competes with other demands such as food production and personal care products, which can result in higher prices due to limited supply. Additionally, the process of cultivation, harvesting, and processing these vegetable oils. can be resource-intensive, involving land, water, and energy, contributing to production costs [1, 22].

Furthermore, there is a loss of atomic efficiency in the production process; during the production of HVO, oxygen is removed from the triglyceride in the form of CO_2 and H_2O, resulting in a smaller amount of product obtained for the same amount of raw material used [15]. Moreover, the process consumes H_2, typically derived from non-renewable fuels, which contributes to increasing production costs. This need for additional resources can make HVO less competitive compared to other renewable energy sources unless more efficient and economical ways of producing hydrogen are developed, such as using alternative sources like electrolysis or biomass [23].

4 Environmental Impact

In 2019, total CO_2 emissions from fossil fuel sources reached approximately 37 billion tons, with oil contributing about 13 billion tons to this number [24]. The combined consumption of diesel and aviation fuel reached approximately 36 million barrels per day, which is about 36% of the total oil consumed. This proportion potentially resulted in the emission of about 5 billion tons of CO_2. In the face of this alarming reality, to combat these numbers, a thorough analysis of the life cycle of HVO is necessary as a crucial measure, especially for the heavy diesel-powered transport industry, for which there are currently few alternatives for emission reduction [25].

Life Cycle Assessment (LCA) is an essential tool for understanding the environmental impacts associated with a product, process, or activity throughout all its stages, from raw material extraction to final disposal. In the context of biofuels, such as hydrotreated vegetable oil, LCA plays a crucial role in assessing the potential environmental benefits and challenges associated with its production and use. This analytical method allows for the identification and quantification of environmental impacts throughout the entire life cycle, from production to use [26].

Various studies have addressed the life cycle assessment of renewable diesel, especially in contexts of cogeneration of energy, using vegetable oils from various sources. In these analyses, various environmental impact categories were considered, including global warming potential, acidification, and eutrophication. A detailed life cycle analysis revealed that hydrotreated vegetable oil (HVO) has a lower global warming potential than conventional diesel, regardless of the raw material used. However, it was identified that soil pollution, one of the main impacts associated with HVO, usually occurs due to microbial activity or material leaks. The global warming potential, in particular, is strongly influenced by nitrous oxide emissions from the soil, while nitrogen oxide and ammonia emissions contribute significantly to the potential for eutrophication and acidification [27].

For a more quantitative analysis, studies indicate different emission intensities for renewable diesel from different sources. Generally, it is observed that HVO from agricultural residues has an emission intensity of around 16 g of CO_2/MJ, while that from forest residues has an intensity of about 31 g of CO_2/MJ [28]. Regarding HVO derived from reused oil, studies show that it can have an emission of 12.2–16.9 g of CO_2/MJ, representing a reduction of 82 to 87% compared to common diesel. The advanced biofuel derived from animal fat can have an emission of approximately 33.3 g of CO_2eq/MJ, resulting in a 65% reduction compared to fossil diesel [29].

Considering these values, the average CO_2 emissions for the use of HVO are estimated to be around 27 g of CO_2/MJ, which represents a significant reduction of approximately 71% in greenhouse gas emissions compared to fossil diesel. This considerable decrease in emissions is crucial to encourage the shift from traditional fuels. Moreover, considering the typical energy content of renewable diesel, which is about 34.4 MJ/l, we can calculate that the average CO_2 emission per liter of HVO is approximately 0.929 kg of CO_2/l. This analysis further reinforces the fundamental role of HVO in reducing carbon emissions and transitioning to a more sustainable energy future. In a global context of increasing environmental awareness and pressure for sustainable solutions, green diesel emerges as a promising alternative, capable of significantly reducing greenhouse gas emissions and contributing to a more ecologically balanced future [24].

5 Future Perspectives

According to the data presented, it becomes evident that hydrotreated vegetable oils are one of the possible ways to increase the index of renewable energy worldwide, which would provide an option to meet the CO_2 reduction target prescribed in the REPowerEU Plan [30]. At the same time, it will be necessary to include these advanced biofuels in legislation and establish clear rules for their use [22]. In addition to these changes, alterations in the composition of diesel in the future can also be expected. HVO is still a new and relatively unknown biofuel, lacking a more developed commercial distribution network [31].

The introduction of changes that have a positive effect on reducing harmful emissions and particulate matter pollution will imply an increase in production costs, and therefore, the speed of their implementation will depend on the economic situation and legislative changes adopted by countries. Regarding its technological development, the issue of advanced biofuels is only at an initial stage. The most commonly declared "first generation" of biofuels is bioethanol produced from starch and sugar, biodiesel produced from vegetable oils (rapeseed, soy, etc.) and animal fats without chemical treatment. These are sophisticated technologies and, above all, commercially available, which are widely studied and of significant global importance currently [32].

Today, there is an HVO production capacity already under construction that will take the possibility of fuel production to 14.4 billion gallons by 2025, and there are possible mapped projects that could take this capacity to 32 billion liters [33]. According to IEA data (2021), in 2020, the global consumption of renewable diesel was 0.155 million barrels per day, equivalent to 9 billion liters, representing only

0.5% of the total diesel consumption worldwide. In the coming years, an accelerated growth in demand for this biofuel is expected, driven mainly by initiatives such as the Low Carbon Fuel Standard (LCFS) in the United States, Canada, Europe, and by the increasing global demand for Sustainable Aviation Fuels (SAF) [24].

In parallel to this, it is essential to intensify research incentives and promote a more comprehensive expansion of knowledge in this area. The application of vegetable oils as a renewable raw material has diversified in the most different areas of knowledge. A study analyzed the market trend of vegetable oils in biofuels and showed that, according to research, the "chemical" area predominates for the species generally used, which means that vegetable oils have had an industrial application concentrated in oleochemistry, and it is noticed that the volume of patent documents in the "energy and fuels" area is still not very representative, as shown in table 2. This fact may be related to the level of maturity of the sector under study regarding the application of vegetable oils [31].

However, there are challenges to be overcome for a significant increase in the production of this biofuel, mainly the agronomic aspects and the physicochemical characteristics of vegetable oils to make them suitable for the processes of obtaining biodiesel [31, 34].

Table 2 Documents retrieved by the research of Santos et al. [22], associated with the technological area of energy and fuels, in descending order

Oilseed species	Documents between 2007 and 2018	Area: energy and fuels/chemistry (percentage of total)
Physic nut	2.975	467 (50%)
Macaw palm	50	6 (38%)
Camelina	1.140	107 (25%)
Crambe	612	45 (25%)
Babassu	1.022	79 (22%)
Oil Palm	12.991	681 (16%)
Cotton seed	23.906	1.050 (14%)
Rapeseed	44.739	1.558 (11%)
Castor	72.411	2.375 (11%)
Coconut	40.778	1.096 (8%)
Sunflower	42.763	940 (6%)
Soy	179.490	2.691 (5%)
Peanut	53.781	761(4%)
Flaxseed	7.768	101 (4%)
Corn	111.738	1.322 (4%)
Olive	46.055	623 (4%)
Sesame	57.633	324 (2%)
Ouricury palm	8	0 (0%)

6 Conclusion

In the current context of political and economic crises, the complexity of challenges and collective needs surpass the response capacity of governments. As we face these adversities, it is crucial to recognize that innovative and sustainable solutions are necessary to drive progress and promote social and environmental well-being. In this sense, hydrotreated vegetable oil (HVO) emerges as a fundamental element in the search for cleaner and more efficient energy alternatives. Through its ecologically conscious production and environmentally beneficial properties, HVO represents a significant step towards a more sustainable energy future.

Throughout this analysis, we have explored in detail the potential of this biofuel as a vital component in the transition to a low-carbon economy. Its tangible benefits, which include the reduction of greenhouse gas emissions and the decrease in dependence on fossil fuels, offer a promising perspective for the renewable energy sector. However, for this vision to become a reality, it is imperative to address the challenges inherent in the large-scale adoption of HVO.

The growth of the biofuel market presents significant opportunities but also implies facing issues such as competition with other renewable energy sources and the need for adequate infrastructure. Furthermore, proactive government policies and substantial investments are essential to stimulate innovation and create an environment conducive to the development and adoption of HVO. Therefore, it is crucial that various sectors of society work together to promote the advancement of green diesel and ensure a more sustainable energy future for future generations.

References

1. Glisic SB, Pajnik JM, Orlovic AM (2016) Process and techno-economic analysis of green diesel production from waste vegetable oil and the comparison with ester type biodiesel production. Appl Energy 170:176–185
2. Ritchie H, Rosado P. Fossil fuels. Our World in Data. https://ourworldindata.org/fossil-fuels. Last Accessed 26 March 2024
3. Zhou Y, Pavlenko N, Rutherford D, Osipova L, Comer B (2020) The potential of liquid biofuels in reducing ship emissions. Retrieved from the International Council on Clean Transportation
4. MIT Technology Review, Do Biofuels Reduce Greenhouse Gases? https://www.technologyreview.com/s/609019/the-dark-secret-at-the-heart-of-ai/. Last Accessed 26 March 2024
5. Ministério de Minas e Energia (2020). https://antigo.mme.gov.br/web/guest/secretarias/petroleo-gas-natural-e-biocombustiveis/acoes-e-programas/programas/biodiesel/perguntas-frequentes. Last Accessed 26 March 2024
6. Soriano JA, Agudelo JR, López AF, Armas O (2017) Oxidation reactivity and nanostructural characterization of the soot coming from farnesane—A novel diesel fuel derived from sugar cane. Carbon 125:516–529
7. Serrano-Ruiz JC, Ramos-Fernández EV, Sepúlveda-Escribano A (2012) From biodiesel and bioethanol to liquid hydrocarbon fuels: new hydrotreating and advanced microbial technologies. Energy Environ Sci 5(2):5638–5652
8. Racaczeski CC, Zancanaro D, Alzani A, Ferreira EA (2006) Biodiesel, em combustível em expansão. Synergismus scyentifica UTFPR, Pato Branco, 1(1–4):324–330

9. Sousa FP, Silva LN, de Rezende DB, de Oliveira LCA, Pasa VM (2018) Simultaneous deoxygenation, cracking and isomerization of palm kernel oil and palm olein over beta zeolite to produce biogasoline, green diesel and biojet-fuel. Fuel 223:149–156
10. Dimitriadis A, Natsios I, Dimaratos A, Katsaounis D, Samaras Z, Bezergianni S, Lehto K (2018) Evaluation of a hydrotreated vegetable oil (HVO) and effects on emissions of a passenger car diesel engine. Front Mech Eng 4:7
11. Douvartzides SL, Charisiou ND, Papageridis KN, Goula MA (2019) Green diesel: biomass feedstocks, production technologies, catalytic research, fuel properties and performance in compression ignition internal combustion engines. Energies 12(5):809
12. Sonthalia A, Kumar N (2019) Hydroprocessed vegetable oil as a fuel for the transportation sector: a review. J Energy Inst 92(1):1–17
13. Rocha DHD, de Souza TAZ, Coronado CJR, Silveira JL, Silva RJ (2021) Exergoenvironmental analysis of hydrogen production through glycerol steam reforming. Int J Hydrogen Energy 46(1):1385–1402
14. Kochetkova D, Blažek J, Šimáček P, Staš M, Beňo Z (2016) Influence of rapeseed oil hydrotreating on hydrogenation activity of CoMo catalyst. Fuel Process Technol 142:319–325
15. Phimsen S, Kiatkittipong W, Yamada H, Tagawa T, Kiatkittipong K, Laosiripojana N, Assabumrungrat S (2016) Oil extracted from spent coffee grounds for bio-hydrotreated diesel production. Energy Convers Manag 126:1028–1036
16. Calero J, Luna D, Sancho ED, Luna C, Bautista FM, Romero AA, Verdugo-Escamilla C (2015) An overview on glycerol-free processes for the production of renewable liquid biofuels, applicable in diesel engines. Renew Sustain Energy Rev 42:1437–1452
17. Bezergianni S, Dimitriadis A (2013) Comparison between different types of renewable diesel. Renew Sustain Energy Rev 21:110–116
18. Kalnes TN, Koers KP, Marker T, Shonnard DR (2009) A technoeconomic and environmental life cycle comparison of green diesel to biodiesel and syndiesel. Environ Progr Sustain Energy: Offic Public Am Inst Chem Eng 28(1):111–120
19. Simacek P, Soucek I, Pospisil M, Vrtiska D, Kittel H (2019) Impact of hydrotreated vegetable oil and biodiesel on properties in blends with mineral diesel fuel. Therm Sci 23(Suppl. 5):1769–1777
20. Hilbers TJ, Sprakel LM, van den Enk LB, Zaalberg B, van den Berg H, van der Ham LG (2015) Green diesel from hydrotreated vegetable oil process design study. Chem Eng Technol 38(4):651–657
21. Mittelbach M (2015) Fuels from oils and fats: Recent developments and perspectives. Eur J Lipid Sci Technol 117(11):1832–1846
22. dos Santos AC, dos Anjos SSN, Braga M, Viana NM, Soares IP (2022) Aplicação industrial de óleos vegetais em biocombustível: um estudo prospectivo e análise de sinais para apontar tendências de mercado. Cadernos de Ciência & Tecnologia 39(2):26995
23. Brannstrom C, Gorayeb A, de Sousa Mendes J, Loureiro C, de Andrade Meireles AJ, da Silva EV, de Oliveira RF (2017) Is Brazilian wind power development sustainable? Insights from a review of conflicts in Ceará state. Renew Sustain Energy Rev 67:62–71
24. IEA (2021) World Energy Outlook 2021. Paris: IEA. https://www.iea.org/reports/world-energy-outlook-2021. Last Accessed 26 March 2024
25. OPEC (2021) World Energy Outlook. https://www.opec.org/. Last Accessed 26 March 2024
26. Hunt R, Franklin E (1996) LCA-How it came about. Personal reflections on the origin and the development of LCA in the USA. Int J LCA 1
27. Arvidsson R, Persson S, Fröling M, Svanström M (2011) Life cycle assessment of hydrotreated vegetable oil from rape, oil palm and Jatropha. J Clean Prod 19(2–3):129–137
28. Christensen A (2021) Transportation carbon intensity targets for the European Union—Road and aviation sectors
29. Seber G, Malina R, Pearlson M, Olcay H, Hileman J, Barrett S (2014) Environmental and economic assessment of producing hydroprocessed jet and diesel fuel from waste oils and tallow. Biomass Bioenergy 67

30. European C (2018). https://eur-lex.europa.eu/legal-content/EN/TXT/?uri=CELEX%3A52018DC0757. Last Accessed 26 March 2024
31. dos Santos AC, Ferreira PM, Lopes CL, Braga M, Viana NM (2022) Estudo prospectivo de óleos vegetais
32. Asmelash E, Gorini R (2021) International oil companies and the energy transition. Int Renew Energy Agency (IRENA), Abu Dhabi
33. Stanley M (2020) A fuel for transition: the renewable diesel opportunity
34. Zeman P, Hönig V, Kotek M, Táborský J, Obergruber M, Mařík J, Hartová V, Pechout M (2019) Hydrotreated vegetable oil as a fuel from waste materials. Catalysts 9:337

Brewer's Spent Grain Valorization to Reduce the Energy Bill of Small Craft Breweries in Portugal

L. Rodrigues, C. M. Rangel, R. Panízio, G. Lourinho, and R. Traquete

Abstract Beer is one of the most consumed drinks around the world. Beer production, with a significant consumption of energy, mostly of fossil origin, generates large amounts of waste, raising some concerns due to its environmental impact. Many organic wastes can be converted into value-added products. In particular, the cereal bagasse resulting from the beer production process, Brewer's Spent Grain (BSG), is proposed as a substrate in anaerobic digestion (AD) to produce biogas and biomethane. The liquid and solid digestates resulting from this process can be used as biofertilizers. Alternatively, the biogas and the liquid portion of the digestate can be processed to obtain hydrogen through reforming and electrolysis, respectively, which can also be used to generate heat or electricity to meet the energy needs of the breweries themselves. Wastewater from the brewing process or from other productions neighboring the factory, such as from olive oil mills or cheese factories, can be co-processed by AD and serve as feed for electrolysis to produce hydrogen, contributing to the treatment and promotion of the circularity of water as a resource.

Keywords Brewer's spent grain · Anaerobic digestion · Biomethane · Green hydrogen

L. Rodrigues
Instituto Politécnico de Portalegre, Portalegre, Portugal
e-mail: luiz.rodrigues@ppt.pt

C. M. Rangel (✉)
Laboratório Nacional de Energia e Geologia, Lisboa, Portugal
e-mail: carmen.rangel@lneg.pt

R. Panízio
Valoriza—Centro de Valorização de Recursos Endógenos, Portalegre, Portugal

G. Lourinho
BIOREF—Collaborative Laboratory for Biorefineries, Portalegre, Portugal

R. Traquete
Barona Brewing Company Lda, Marvão, Portugal

P. S. Brito et al. (eds.), *Insights into Energy Trends*, Green Energy and Technology,
https://doi.org/10.1007/978-3-031-83811-8_11

1 Introduction

Beer production is associated with the generation of large volumes of residual biomass. Various wastes are generated in the brewing process, including liquid and solid effluents, in particular malt bagasse, also known as Brewer's Spent Grains (SBG) (Fig. 1).

BSG is a perishable product due to its high moisture content, representing ~80 to 85% of total solid waste from breweries [1–4], which causes environmental concerns in virtue of its high organic load. Thus, the recent spread of craft breweries across the Portuguese territory brings about important challenges and opportunities for the treatment and applications of the mentioned waste, in the spirit of the circular bioeconomy and to face the water-energy nexus [1, 4].

For every 1000 t of beer produced, 137–173 t of waste is generated [5], leading to 20–28 kg of wet BSG for every hectoliter (hl) of beer obtained [1, 5, 6]. In the period 2008–2021, beer production in Portugal was, on average, 7.3×10^6 hl /year [7]. After a slight decline in 2020, due to the pandemic context, in 2021 production reached 6.7×10^6 hl, a slight increase compared to the previous year [7], corresponding to ~134,000 t/year of wet BSG. In 2021, around 100 producers were registered in Portugal, 98 of them microbreweries [8]. This proliferation and the consequent generation of brewing waste throughout the country, predicts additional environmental load in locations with limited capacity to absorb this type of waste.

The main destination of BSG has been animal feed [9] or disposal in landfills, with costs for brewers and the environment, in addition to the waste of an important resource. Since BSG is rich in fiber, protein, minerals, and vitamins it is very attractive as a dietary supplement in these contexts [5]. The dry BSG may contain, among others, around 12–25% cellulose, 20–25% hemicellulose, 12–28% total lignin, in addition to ~10% lipids, ~ 8% extractable materials and 2–5% ash [5, 6].

This relatively high content of lignocellulosic fibers and proteins points to applications such as the production of 2nd generation bioethanol, hydrothermal coal, lactic acid, hydroxymethyl furfural and products intended for human nutrition, among others [4].

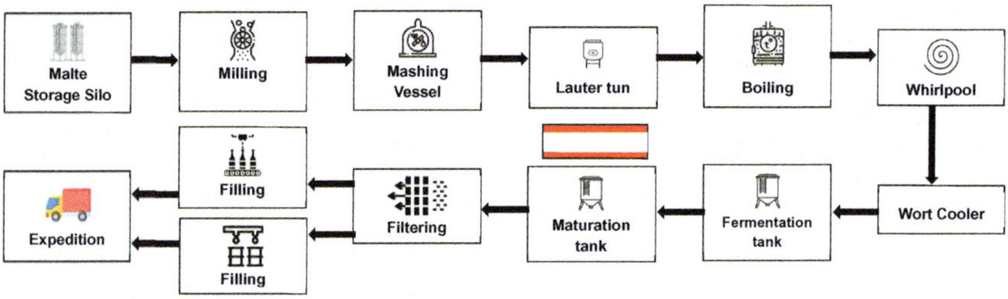

Fig. 1 Schematic representation of the brewing process, with the BSG generation stage highlighted in red [1]

In fact, the energy recovery of BSG has been the most sought-after objective, including the production of biofuels such as bioethanol, biobutanol, biohydrogen, biodiesel and biogas [9, 10]. Biogas is the biofuel that has received the most attention [1, 11–13]. This is due to the relatively high levels of volatile materials, low fixed carbon, and values of the Higher Heating Value (HHV) that vary between 11.1 and 21.0 MJ/kg [14], making BSG an interesting energy resource to cover, at least partially, the electrical and thermal energy needs of breweries [9, 13, 15].

In this work, Brewer's Spent Grain (BSG), is proposed as a raw material to be used as a substrate in anaerobic digestion (AD) to produce biogas. The production of hydrogen will be also assessed in order to generate heat/electricity to meet some of the energy needs of the brewery, through reforming and electrolysis that can be implemented, respectively, by processing the produced biogas and the liquid portion of the digestate.

Proximate, Ultimate, and Thermogravimetric analysis of the BSG sample is presented in Table 1. Selected BSG contained, in terms of its elemental composition, approximately 45 wt% of C, 45 wt% of O, 7 wt% H, 3 wt% N and negligible amounts of S. The sample presented around 7 wt% of moisture content, a very high content of volatiles and low ash, representing 83 and 3 wt%, respectively.

Figure 2 shows TGA data and mass loss rate profile of the selected BSG, run at 30 °C/min, taken in order to assess its thermal stability. Three different regions can be identified. The first peak of the DTA curve, running from around 50 to 180 °C, seems to be related with some residual moisture present in the analyzed samples [16]. The 2nd peak centered around 320 °C may be attributed to the degradation of cellulose and hemicellulose, whereas the breakdown of lignin appears to occur at temperatures around 600 °C [16].

A kinetic model will be fitted to data in order to assess transitions for applicable waste conversion methods for energy and material recovery considered crucial for sustainable circular bioeconomy.

Figure 2 compares quite well with results presented in the literature by other authors [17, 18], using spent grain for thermochemical conversion, representing a promising feedstock for further development of renewable and sustainable chemicals and fuels.

Table 1 Proximate and ultimate analysis of a BSG sample supplied by a microbrewery located in North Alentejo region of Portugal

Proximate analysis (%)				Ultimate analysis (%)				
Humidity*	Ashes	Volatile matter	Fixed carbon	C	H	N	O	S
7.20	2.88	83.03	14.09	45.13	6.69	3.21	44.97	0.00

*72% wet basis as received

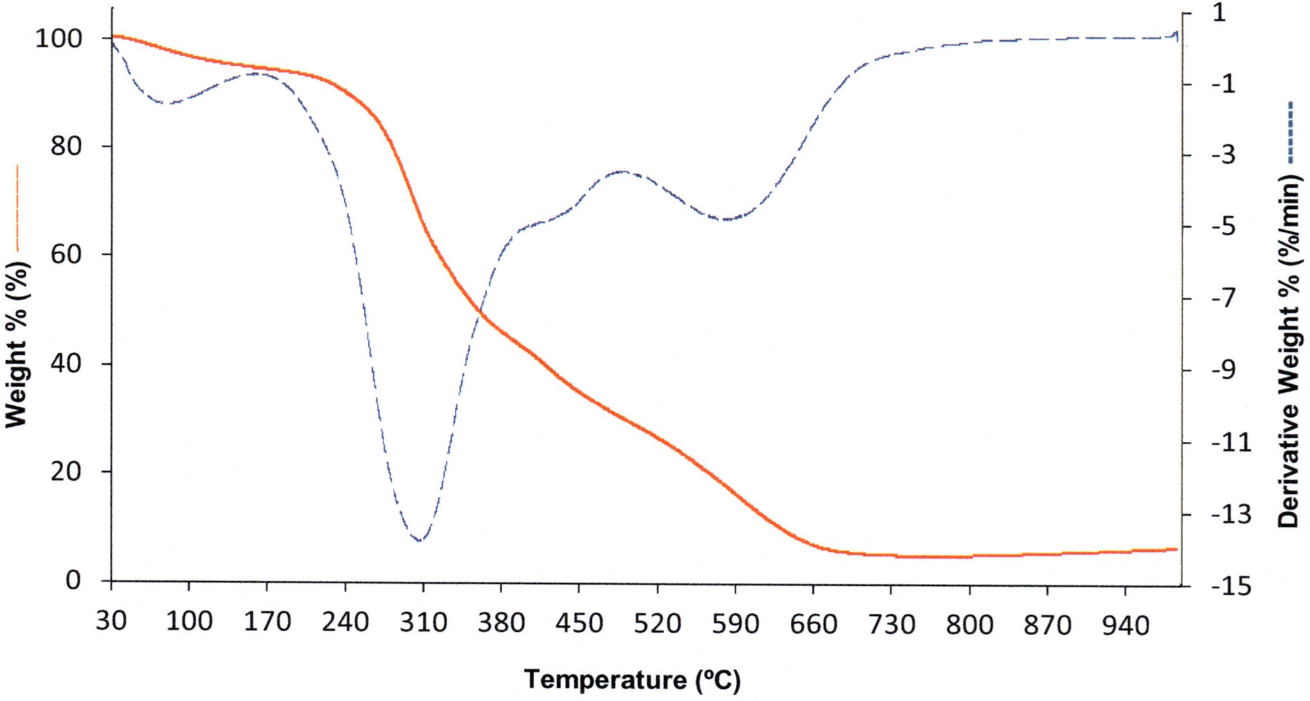

Fig. 2 Thermogravimetric (red solid line) and derivative of thermogravimetric curves (blue dash line) of the studied BSG sample supplied by a microbrewery located in North Alentejo region of Portugal, performed at 30 °C/min temperature change rate

2 Biogas Generation Through Anaerobic Digestion (AD)

Biogas is the product of the anaerobic decomposition (AD) of organic matter, by the action of a consortium of microorganisms, the majority of which consists of methane (55–70%), CH_4, and carbon dioxide, CO_2 (30 to 45%) [13, 19]. The composition of biogas may vary depending on the composition and pre-treatment routes of the biomass used as substrate for the microorganisms, as well as operational parameters of the AD, such as the pH value and C/N ratio, among others. Biogas may also contain other gases such as N_2, H_2, O_2, H_2S, NH_3 and hydrocarbons, in significant quantities that may affect, to a greater or lesser extent, the conduction and performances of the AD process or the further usage of the obtained biogas [9].

Biogas can be purified to obtain biomethane, with greater heating value, and may fully replace natural gas in the generation of thermal and electrical energy. Furthermore, biomethane can be further transformed into H_2 through water vapor reforming, with additional environmental advantages.

Biomethane biogas purification may not be such an accessible operation for micro-breweries, given the still high costs of commercially available technologies. Direct burning of biogas to generate heat could be a more viable alternative for craft brewers. Breweries are energy intensive, consuming 240–280 kWh of heat and 75–138 kWh of electrical energy per m^3 of produced beer [15].

3 Obtaining Biogas and Biofertilizers from Brewery Waste

The production of biogas through anaerobic digestion (AD) seems to be the most obvious option for energy recovery from BSG, not only due to the favorable chemical composition of this residue, but also given its high moisture content, as generated, of ~70% to 85% [9], well within the values found in the BSG sample used in this work. Other waste energy recovery technologies currently available on the market, especially those based on thermal processes, are less attractive as they require relatively large amounts of energy to pre-dry the biomass.

In the case of AD, the additional water to achieve the necessary anaerobic conditions in the biodigester may even come directly from the beer manufacturing process, which generates large volumes of wastewater [15]. In fact, wastewater resulting from beer manufacturing, with its high levels of organic matter and suspended solids [16], as well as sludge from brewery wastewater treatment plants (WWTP) can be used to generate biogas and methane [10, 11]. Other waste streams with high levels of organic matter that require treatment before being discharged into the environment, such as effluents from pig farming and cheese factories [10], preferably from neighboring factories, could also help meet the need for water, with advantages for circularity of this increasingly scarce resource in Portugal, and elsewhere.

Another advantage of the AD of brewer's malt bagasse is that, in addition to the biogas obtained, which can be burned directly to generate the heat necessary for boiling the wort and other operations, the liquid and solid digestates, that result from the process, can be used directly as biofertilizers or soil conditioners in agriculture or gardening.

4 Possibilities for Obtaining Hydrogen from Brewing Waste

In many energy transition scenarios, hydrogen (H_2) has been identified and proposed as an important carrier of energy, raw material, and fuel for use in heat, electricity, and steam generation systems, as well as in road, rail, and maritime transport systems, due to its high heating value and ecological potential, given that its combustion only produces water as byproduct.

It was already mentioned that, in a beer factory, H_2 could be produced by steam reforming of the biogas originated from the BSG. Lately, some attention has also been paid to the production of biohydrogen from brewing waste, particularly wastewater, using various strains of hydrogenogenic bacteria [20].

In breweries, H_2 can also be obtained from the BSG via the electrolysis of the liquid digestate resulting from its AD or from co-digestion with other wastes from inside or outside the factory, such as wastewater from the factory's or neighboring WWTPs, olive oil mills or cheese factories.

The electrolysis of wastewater generated in the brewing process to obtain hydrogen has been studied [21, 22]. Although, commercial electrolytic production of H_2 uses high purity water, there are indications that the presence of substances such as cellulose and lignin, as in brewing water and wastewater, can contribute to the reduction of the electrical potential of H_2 discharge, and thus to improve the energy efficiency of the process [23, 24]. But there are other environmental and economic advantages to wastewater electrolysis, particularly if green electricity is used. These include the simultaneous oxidation of toxic or recalcitrant components, the electro-coagulation of fine particles, as well as the possibility of producing valuable products, and storing excess renewable energy in the form of green H_2 [23].

There is also the possibility of injecting the H_2 produced by electrolysis into the biodigester to stimulate the biological conversion of CO_2 into CH_4, increasing its content in the biogas obtained, and thus also increasing its heating value [25].

Solid oxide electrolyzers (SOE) could be very attractive in the context of wastewater use, since water steam production integrates the function of preliminary water treatment [22]. Its potential and viability are being examined given its high efficiency. Solid oxide cells are intrinsically reversible allowing operation in electrolysis or fuel cell mode, for hydrogen and electricity production, respectively, thus only once device is needed [22, 26].

Table 2 Available energy/needs for the considered brewery according to different authors

BSG (kg/day)	Generated biogas (m^3)	Available energy (kWh)	Energy needs_ Heat (kWh)/ coverage** (%)	Energy needs_ Electricity (kWh)/ coverage** (%)
100	13.4 [27]	59.5*	120–140 [15]/ ≥34	38–69 [15]/≥26
	7.2 [28]	32.1*	≥18	≥14

*Considering an average calorific value of 16 MJ/m^3 according to IEA [29]
**Considering average 80% and 30% efficiency, respectively, for heat and electricity conversions

5 BSG Energy Recovery Potential. Final Comments

Considering the size of craft breweries in Portugal, this brief analysis points to the potential for energy recovery from BSG, proposing the integration of AD to produce biogas and biofertilizers, with the electrolysis of effluent waters from this process with that of brewing itself, to produce hydrogen H_2 and achieve water treatment. This proposal seems very attractive and accessible for craft beer factories, given the availability and maturity of the technologies involved, with several alternatives emerging for the use of the produced gases [19]:

(a) Burning of biogas or CH_4 as fuel in boilers to meet the heat needs of manufacturing and hygiene requirements (brewing, fermentation, storage, bottling and barreling, boilers, compressors, and cleaning tasks) [19]. The most energy consuming stages are heating and boiling the wort and pasteurization, although the heat can be recovered.
(b) Conversion of biogas into H_2 through reforming for subsequent use in fuel cells. This alternative, although attractive from the viewpoint of energy efficiency, may be financially prohibitive, due to the high investments needed for the equipment required for the purification of the gas resulting from the reforming [19].
(c) Burning of mixtures of methane with hydrogen ("hythane"), a fuel with a high heating value that would allow heat to be obtained for boilers, without large additional investments or more stringent safety measures than those in force for gaseous traditional fuels. since the H_2 content in hythane is typically low [19].

In the context of craft breweries in Portugal, taking as a reference a manufacturing unit producing 500 L beer/day (20 kg BSG /hl), it is estimated that 13.4 m^3 of biogas will be generated, considering 57% of biomethane in the biogas [27]. The available energy in the form of heat or electricity will be able to cover the brewery needs in ~34% or 26% respectively, see Table 2. In the consideration of these values, caution must be taken since it may imply significant, non affordable, capital costs investments in the case of such small breweries, to reappraise when performing the indispensable economic feasibility studies.

Acknowledgements Acknowledgements are due to Barona Brewing Company, Lda., Marvão, Portugal, for providing the BSG samples.

References

1. Pabbathi NPP, Velidandi A, Pogul S, Gandam PK, Baadhe RR, Sharma M, Sirohi R, Thakur VK, Gupta VK (2022) Brewer's spent grains-based biorefineries: a critical review. Fuel 317:123435
2. Emmanuel JK, Nganyira PD, Shao GN (2022) Evaluating the potential applications of brewers' spent grain in biogas generation, food and biotechnology industry: a review. Heliyon 8:e11140
3. Mitri S, Salame S-J, Khelfa A, Leonard E, Maroun RG, Louka N, Koubaa M (2022) Valorization of Brewers' spent grains: pretreatments and fermentation, a review. Fermentation 8:50
4. Zeko-Pivač A, Tišma M, Žnidaršič-Plazl P, Kulisic B, Sakellaris G, Hao J, Planinić M. The potential of brewer's spent grain in the circular bioeconomy: state of the art and future perspectives. Front Bioeng Biotechnol 10:870744. https://doi.org/10.3389/fbioe.2022.870744
5. Lynch KM, Steffen EJ, Arendt EK (2016) Brewers' spent grain: a review with an emphasis on food and health. J Inst Brew 122:553–568
6. Massardi MM, Massini RMM, Silva DJ (2020) Caracterização química do bagaço de malte e avaliação do seu potencial para a obtenção de produtos de valor agregado. J Eng Exact Sci 6. https://doi.org/10.18540/jcecvl6iss1pp0083-0091
7. Statista: Volume of beer produced annually in Portugal from 2008 to 2021, https://www.statista.com/statistics/446698/volume-beer-production-portugal/. Last Accessed 26 Nov 2023
8. Cervejeiros de Portugal: Dados Económicos. https://www.cervejeirosdeportugal.pt/dados_economicos/. Last Accessed 10 Oct 2023
9. Glowacki S, Salamon A, Sojak M, Tulej W, Brys AJ, Hutsol T, Salamon M, Kukharets S, Janaszek-Mankowska M (2022) The use of brewer's spent grain after beer production for energy purposes. Materials 15:3703
10. Puligundla P, Mok C (2021) Recent advances in biotechnological valorization of brewers' spent grain. Food Sci Biotechnol. 30:341–353
11. Zupancic GD, Panjicko M, Zelic B (2017) Biogas production from brewer's yeast using an anaerobic sequencing batch reactor. Food Technol Biotechnol 55:187–196
12. Endler IA (2021) Avaliação da produção de biogás a partir de resíduo sólido orgânico da indústria cervejeira. Universidade do Vale de Taquari, Brasil
13. de Amorim HJCAL (2022) Recuperação de energia a partir do tratamento de resíduos de uma indústria cervejeira artesanal. Universidade Federal de Alagoas, Brasil
14. Cordeiro LG (2011) Caracterização e viabilidade económica do bagaço de malte oriundo de cervejarias para fins energéticos. Universidade Federal da Paraíba, Brasil
15. Rawalgaonkar D, Zhang Y, Walker S, Kirchman P, Zhang Q, Ergas SJ (2023) Recovery of energy and carbon dioxide from craft brewery wastes for onsite use. Fermentation 9:831
16. Grotto CGL, Colares CJG, do Vale AT, Pereira DH (2021) Energy potential of spent grains from brown porter style beer production. Energia na Agricultura 36(2):204–217
17. Araújo TP, Quesada B, Bergamasco R, Vareshini DT, Barros MASD (2020) Activacted hydrochar produced from brewer's spent grain and its application in the removal of acetaminoohen. Bioresour Technol 310:123399
18. Sobek S, Zeng K, Werle S, Junga R, Sajdak M (2022) Brewer's spent grain pyrolysis kinetics and evolved gas analysis for the sustainable phenolic compounds and fatty acids recovery potential. Renew Energy 199:157–168
19. Arantes MK, Alves HJ, Sequinel R, da Silva EA (2017) Treatment of brewery wastewater and its use for biological production of methane and hydrogen. Int J Hydrogen Energy 42:26243–26256
20. Soares JF, Mayer FD, Mazutti MA (2024) Hydrogen production from brewer's spent grain hydrolysate by dark fermentation. Int J Hydrog Energy 52:352–363
21. Orosa LP, Chinarro E, Guinea D, García-Alegre MC (2022) Hydrogen production by wastewater alkaline electro-oxidation. Energies 15:5888
22. Baldinelli A, Cinti G, Barelli L, Bidini G (2022) Hydrogen production from low quality water: challenges and perspectives. J Phys: Conf Ser 2385:012048

23. Luo H, Barrio J, Sunny N, Li A, Steier L, Shah N, Stephens IEL, Titirici M-M (2011) Progress and perspectives in photo- and electrochemical oxidation of biomass for sustainable chemicals and hydrogen production. Adv Energy Mater 11:2101180
24. Dolle C, Neha N, Coutanceau C (2022) Electrochemical hydrogen production from biomass. Curr Opin Electrochem 31:100841
25. Zhang L, Kuroki A, Tong YW (2020) A mini-review on in situ biogas upgrading technologies via enhanced hydrogenotrophic methanogenesis to improve the quality of biogas from anaerobic digesters. Front Energy Res 8:69
26. Maddaloni M, Marchionni M, Abbá A, Mascia M, Tola V, Carpanese MP, Bertanza G, Artioli N (2023) Exploring the viability of utilizing treated wastewater as a sustainable water resource for green hydrogen generation using Solid Oxide Electrolysis Cells (SOECs). Water 15:2569
27. Onudi et al (2022) Biogás no setor cervejeiro: nota técnica. Brasília: MCTI, E-book
28. Oliveira JV, Alves MM, Costa JC (2018) Biochemical methane potential of brewery by-products. Clean Technol Environ Policy 20:435–440
29. Outlook for Biogas and Biomethane—Prospect for Organic Growth. IEA (2020). https://iea.blob.core.windows.net/assets/03aeb10c-c38c-4d10-bcec-de92e9ab815f/Outlook_for_biogas_and_biomethane.pdf

Application of Renewable Energy—The Case of San Tomé Island

Ludcelma de Ceita, Ricardo Manso, and Teresa Eugénio

Abstract In the current context, the energy sector is one of the most important strategic points for the economic transformation of any society. Access to energy is a universal human right and a concern of the state. Whatever national development strategy is defined or implemented, it will always require measures to promote access to energy for all. Although the strong growth of renewable energies worldwide is evident, there are still few studies on their applicability in developing countries. The aim of this study is to contribute to energy sustainability in an emerging country where renewable energies have been little analysed and developed. The country analysed is São Tomé and Príncipe (STP). This paper develops a study on the application of renewable energy, energy sustainability and clean electricity generation as challenges of the present and ways forward. Aware of the potential of natural resources on the islands of São Tomé and Príncipe, the study follows the following methodology: first phase: a preliminary study to describe the current energy situation on the island; second phase: presentation of a simulation on the use of available resources that has a direct impact on minimising the consumption of fossil fuels on the islands and on the country's dependence on foreign fuel imports. As a result, the following proposal is made: Utilisation of photovoltaic solar energy with the help of agro-voltaic systems and floating panels, a biomass plant and a transition to electromobility with the help of electric vehicles as energy storage and means of transport.

Keywords Energy sustainability · Emerging countries · São Tomé and príncipe

L. de Ceita · R. Manso
School of Technology and Management, Polytechnic of Leiria, Leiria, Portugal

T. Eugénio (✉)
CARME—Centre for Applied Research in Economics and Management, School of Technology and Management, Polytechnic of Leiria, Leiria, Portugal
e-mail: teresa.eugenio@ipleiria.pt

P. S. Brito et al. (eds.), *Insights into Energy Trends*, Green Energy and Technology,
https://doi.org/10.1007/978-3-031-83811-8_12

1 Introduction

Access to modern energy is one of the most important items on the agenda of the Sustainable Development Goals (SDGs) and the African Union Commission. Investments in energy are a contribution to poverty reduction and to supporting the economy and sustainable growth in Africa. It must be reliable and affordable to ensure the use of clean and high-quality electrical services [1–3].

For this gradual transformation to take place, there is an urgent need for strong government leadership, effective planning and clear allocation of responsibilities at national and regional levels to overcome the persistent problems of low access to electricity in Africa and in São Tomé and Príncipe (STP) in particular. The aim of this energy study for STP is to enable this transition. In this context, the case study consists of presenting practical solutions to meet the energy needs of São Tomé and Príncipe in order to end the dependence on external fuel imports and generate electricity for the communities of the archipelago. Clean, safe and affordable energy production is a key factor in alleviating poverty, securing rural livelihoods, attracting industrial investment to diversify the national economy and connecting the country through trade and co-operation. Renewable energy is abundant in the STP. Their sources have the potential to meet the entire energy demand, facilitating the rapid transition from energy to renewables and improving the country's quality of life and environment. It is clear that there are many challenges to overcome in order to achieve the energy transition in the STP. These challenges include affordable access to modern energy services, minimising continued reliance on traditional energy sources, eliminating dependence on fossil fuels. With one caveat: some fuel will still be needed for lubricants, fishing boats and tractors, as there are no electric vehicles in these categories on the islands yet, as well as for the aircraft that provide inter-island transport.

The success of the energy transition depends on the leadership of the country's government and the prioritisation of the process in the government's plan. This means fully mobilising its own resources, putting the energy sector at the top of the national and regional agenda, and adopting approaches that put the islands directly on innovative and low-carbon energy development pathways. This study argues that the pathway for energy transformation in the country can be defined by a few basic strategies: Building an energy infrastructure adapted to economic and social development, developing the renewable energy sector in line with existing international agreements to take advantage of the large potential for energy from renewable sources, investing in energy efficiency programmes for buildings, industry and transport, and other adaptation strategies that promote this transition. Considering the size of the island, the study was adapted to the transport sector as one of the alternatives for energy storage as well as for the sustainability of the country with the introduction of electric vehicles in it.

The introduction of electric vehicles on the islands would be advantageous, as the island is suitable for this type of transport and batteries with a long autonomy would not be necessary. The possibility of a biomass plant was also analysed, adapting the existing thermoelectric plants for night-time energy production. In order to establish the energy profile of STP, it was possible to calculate the total annual primary energy of the country using reports from the STP Water and Electricity Company (EMAE) and official reports from the island. Based on this information, the annual useful energy for each case study, the energy sector and the road sector could be determined. In this context, this study invites the country to embrace and analyse these renewable opportunities as a way to continuously improve the country's path towards economic autonomy.

2 A Brief Introduction to the Archipelago of São Tomé and Príncipe

The country has business opportunities in various areas such as agriculture and fishing, extractive industries, tourism and services, as well as construction. It is part of the country's strategy to invest in new infrastructure rehabilitation, maintenance and construction projects, energy, roads, ports, airports, agro-tourism, tourism and housing projects, among others, and it is hoped that in the medium term these sectors will contribute more actively to economic growth and diversification. STP's geographical location provides favorable conditions to produce electricity through PV panels. The climate in STP has many days with clouds, which can affect its performance, since it depends on direct solar radiation. In STP, there are 1760 h of sunshine throughout the year and the solar potential in São Tomé and Príncipe is around 4 kWh/kWp7, with vast areas susceptible to the development of solar PV plants, and there is also potential for one-off projects developed more at a rural level, in schools or by private initiative [4].

Biomass is abundant on the island. São Tomé and Príncipe's energy matrix is mostly characterized by biomass consumption. This trend is due to the country's socio-economic characteristics, where a large part of the population uses firewood and coal as a source of domestic energy. Increased urbanization means, among other things, a greater need for transport, and the country's road sector8 is old and deficient. Average traffic is around 1,500 vehicles per day, with a high presence of motorcycles in urban areas. From the capital of São Tomé to the city of Porto Alegre is approximately 71 km, thus representing the longest distance in km in the country; the second longest connection is from the city of São Tomé to the city of São João dos Angolares with around 40 km; and the third longest connection is from the city of São Tomé to the city of Neves with 27 km. More than 50% of the population travel on foot, which is the main way of getting around, followed by paid private motorized transport with 30.4%. Personal transport accounts for 8.4% and public transport for 7.4%. The climate is tropical and humid, with temperatures varying between 22 °C and 32 °C all year round and its mountainous orography providing various microclimates. There are only two climatic seasons, the drier season with a slight drop in temperature, which occurs between June and September, locally known as Gravana, and the rainy season which occurs in the remaining months, with higher temperatures. The higher areas have a high rainfall of up to 7,000 mm per year, while the north and northeast have less rainfall, generally less than 1,000 mm per year.

3 Energy Sustainability—Theoretical Framework and Applied Approach

This chapter provides a theoretical framework for the energy project and its surroundings. The objectives of the project will be presented, giving an overview of the energy sector in STP, as well as identifying and analyzing the current state of the country's energy market. The methodology and vision of the project will be presented, summarizing its key points, motivations, related to the current international context, substantiating that energy is one of the priority areas in the country's development.

Also in this chapter, the assessment of energy production on the islands will be presented, with the aim of profiling the energy consumption of the islands, finding the availability of annual primary energy, as well as their annual energy needs, in order to present a sustainable and renewable energy solution adjusted to their real needs.

Finally, the calculations and results will be presented, with the solution to the project proposal.

Access to clean energy production in STP is low. The central challenge of the project is to explore viable and sustainable solutions, with the overall aim of harnessing existing sustainable resources available on the islands that can be channeled into electricity production and simulating a simple and applicable proposal to respond to issues in energy demand.

Five specific objectives were defined for this work: reducing diesel imports for electricity production; reducing investment in energy infrastructure maintenance; guaranteeing a stable energy service; using local resources for electricity production; encouraging industrial investment in the country.

The project was developed in two phases, in that it began by identifying the problem, surveying and characterizing the availability of primary energy in the country and the hourly/daily electricity consumption on the two islands. A statistical survey was carried out on the amount of fuel imported for energy production and transportation.

After gathering this information about the island, the proposals for producing and improving energy performance were prepared and the results obtained.

Opportunities and threats were identified and analyzed in order to take advantage of existing resources, promoting development that meets the needs of the present without compromising the ability of future generations to meet their own needs.

Current environmental problems, with the identification of global climate change, are the biggest threat that will affect the project's activity. On the other hand, the transformation of the energy sector in STP is a good opportunity to take advantage of for the development of the country, as well as the use of the hydroelectric plants currently decommissioned in the country.

3.1 *Islands, Sustainable Energy—Overview of the Energy Sector in STP*

"Sustainability is the ability to meet our needs in the present without compromising the ability of future generations to meet their own needs. It ensures a balance between economic growth, care for the environment and social well-being",[1] imposing limitations on activities that make it impossible for nature to meet current and future needs. However, it is essential that the country opts for strategies that involve rationalizing and optimizing available resources, creating this sustainable path for the island.

STP's energy sector is 100% publicly owned, and EMAE is the company that currently carries out the activities of production, transmission, distribution and commercialization of electricity, being the main and only one operating in the country today.

In STP, electricity production is based on thermoelectric plants, which use imported diesel. The interconnected electricity system in São Tomé currently has six thermoelectric plants plus one hydroelectric plant, and one thermoelectric plant on Príncipe Island. According to information provided by EMAE, in 2019/2020, the total installed power in the interconnected network in São Tomé is 23,400 MW. The annual energy produced by the hydroelectric plant is 13 GWh[2] and the turbine power is 2 MW.

From the information in the Fig. 1, it can be seen that over the years diesel-based energy production has increased dramatically and investment in hydroelectric energy production has not kept up with this rapid pace until today.[3]

3.2 *Applied Approach*

At this stage of the project, an exhaustive survey was carried out on various essential and relevant pieces of information for its implementation, based on the reality of the country, such as information on radiation, fuel consumption on the islands, information on thermoelectric power stations, information on hydroelectric power stations and information on the transport sector.

Currently, gross annual production at STP is estimated at 110,444,301 kWh and net production or injected into the grid is around 104,952,238 kWh. Total losses are

[1] Source: In "*Our Common Future*" Report by the Brundtland Commission, 1987.

[2] Source: Report: *African energy database base de données energetiques africaines*, 2019.

[3] Source: STP General Directorate of Natural Resources and Energy.

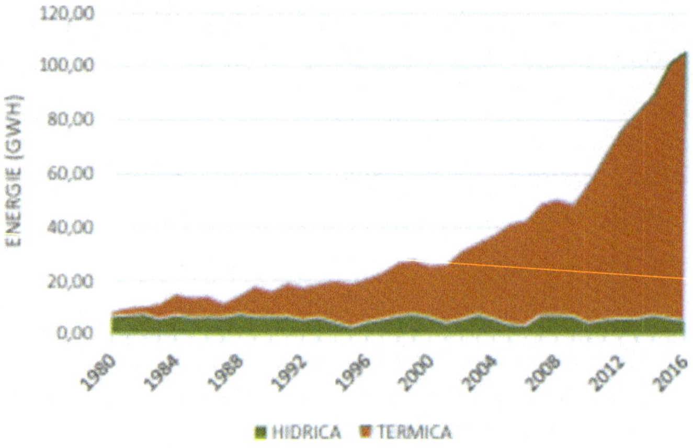

Fig. 1 Evolution of electricity production between 1980 and 2016

28.6%, of which 5.0% represent losses in auxiliary equipment and transformers and the remaining 23.6% are due to commercial losses and transport on the network.

Table 1, which refers to the power stations on São Tomé Island and is supplied by EMAE, contains information on all the generator sets and the number of liters of diesel consumed to run the generators. For its power plants, the company provides a total of 26,193,958 L of diesel per year, with an annual primary energy of 280.7683209 GWh.

From this point, it was possible to draw up a profile of the country's consumption. For calculation purposes, only the annual primary energy was counted, with the help of data provided in reports by international organizations (Table 2).

Based on data from the report "Renewable Energies and Energy Efficiency in São Tomé and Príncipe by the Lusophone Renewable Energy Association (ALER)"[4] it was possible to estimate consumption for the year 2020. The report presents data for 2010, 2011, 2012 and 2014. Based on these figures, consumption was estimated for 2015, 2016, 2017, 2018, 2019 and 2020. Due to the Covid-19 pandemic, consumption in 2019 was assumed to be the same as in 2020. A linear regression was used to find a plausible value for calculation purposes. For 2020 we have a total of 36,225 kt of diesel and 8,238 kt of gasoline. After estimating a value for 2020, the calculation was made for the fuels listed below. For transport, the value presented is equal to the estimated value by subtracting the value of primary energy indicated in the table provided by EMAE, and calculating 421.29675 minus 280.768321, the result being 140.5284 GWh. Primary energy was calculated using the product of its calorific value and its mass (Table 3).

In 2014, primary energy in São Tomé and Príncipe amounted to 72,241.78 toe, broken down into 57.7% fossil fuels, followed by biomass (firewood and charcoal)

[4] Source: National ALER status report, 2020.

Table 1 Diesel consumption of generator sets in interconnected power stations, 2020

Generator set	Annual consumption of diesel (Litres)	Specific consumption g/kWh (annual average)	Standard g/kWh	Density kg/l	Higher calorific values (kWh/Kg)	Annual primary energy (GWh)
	CENTRAL TÉR S. Tomé					
D1	1,403,040,00	294.3	(210–230)	0.846	12.67	15,03,893,321
D3	2,391,360,00	294.2	(210–230)	0.846	12.67	25,6,325,574
ABC3	1,740,738,00	296.7	(210–230)	0.846	12.67	18,65,865,729
CATT		212.5		0.846	12.67	0
PerKis 2	2,115,260,00	437.6		0.846	12.67	22,67,309,119
	CENTRAL TÉR SANTO AMARO 1					
H1	0	0		0.846	12.67	0
H2	2,825,664	255		0.846	12.67	30,2,877,838
H3	2,838,741	255		0.846	12.67	30,42,795,381
H4	2,933,361	255		0.846	12.67	31,44,216,855
H5	2,716,476	255		0.846	12.67	29,11,741,728
	CENTRAL TÉR SANTO AMARO 2					
ABC1	0	0		0.846	12.67	0
ABC2	1,496,259	213.2	190	0.846	12.67	16,03,813,089
ABC3	2,060,770	212.13		0.846	12.67	22,08,902,269
	CENTRAL TÉR SANTO AMARO 3					
ABC 1	0			0.846	12.67	0
ABC 2	1,496,259			0.846	12.67	16,03,813,089
ABC 3	2,060,770			0.846	12.67	22,08,902,269
	CENTRAL DE BBF2					
PERKINS 2	115,260	166.7		0.846	12.67	1,235,451,193
PERKINS 2				0.846	12.67	0
Total (litres/yr)	**26 193 958**					**280,768,321**

with 41.4% and a small share of hydropower corresponding to just 0.9%. With the percentage of 41.4%, the energy available from biomass was estimated.

3.3 Development of the Applied Approach

It is estimated that 26,200 l of diesel were used to produce electricity in 2020, which represents primary energy of 280.8 GWh and production of 97.4 GWh. The

Table 2 Consumption of oil products in kt

Total consumption of petroleum products (kt)											
					2014	2015	2016	2017	2018	2019	2020
					52	55	57	59	61	63	63
Reading	2010	2011	2012	2013	2014	2015	2016	2017	2018	2019	2020
Diesel	23.5	24.7	26,284,72	26	29.9	31,625	32,775	33,925	35,075	**36,225**	**36,225**
Gasoline	6.2	5.5	5874,48	6.2	6.8	7,192,307,692	7,453,846,154	7,715,384,615	7,976,923,077	**8,238,461,538**	**8,238**

Table 3 Energy available in 2020

Primary energy (GWh)			
2020	Total	Thermoelectric	Transport
Diesel	421.3	280.8	140.5
Gasoline	95.8		95.8
Biomass	347.8		

Contador hydroelectric plant supplied the grid with 13 GWh, with a total of 110.4 GWh available on the grid.

The production of 97 GWh is associated with an efficiency of 34.3% of the generator sets, which is reasonable for the Rankine steam cycle. All this production is associated with the aforementioned losses of 28.6%, and the useful electricity consumed by the country is 81.0 GWh. These losses in the network are too high and, for our simulation, we have considered them to be around 10% of electricity production.

Also with the help of EMAE, it was possible to draw up a profile of the country's consumption over the course of a day, as can be seen in the graph below. Figure 2 shows the variation in energy consumption over 24 h. We can see that between 6 p.m. and 9 p.m. peak consumption is reached, which is equivalent to around 19,000 kW.

An estimate was also made of consumption by month, in which we consider that the months from December to March have the highest energy consumption and, conversely, from June to September the lowest. This increase in consumption in the months of December is due to higher temperatures, which justify greater use of refrigeration equipment. During the Gravana season, energy needs are lower.

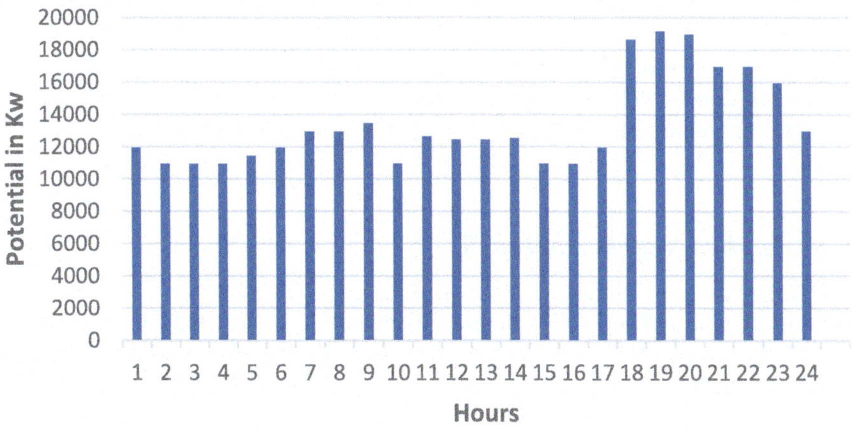

Fig. 2 Load diagram

Table 4 Daily and monthly consumption 2020 (GWh)

Months	Energy per day (GWh)	Expected monthly consumption
January	10 019 200,0	7 223 631,21
February	9 049 600,0	6 524 570,12
March	10 019 200,0	7 223 631,21
April	9 211 200,0	6 641 080,30
May	9 017 280,0	6 501 268,09
June	8 241 600,0	5 942 019,22
July	8 516 320,0	6 140 086,53
August	8 516 320,0	6 140 086,53
September	8 241 600,0	5 942 019,22
October	9 017 280,0	6 501 268,09
November	9 211 200,0	6 641 080,30
December	10 019 200,0	7 223 631,21
		78 644 372
Total	109,08	**78,64,437**
%	**0,720,978,841**	

Table 4 shows[5] the estimated monthly energy requirement, resulting in a total annual requirement of 78.6 GWh, with the 10% losses and without the hydroelectric plant's production.

After analyzing the production and consumption of electricity in STP by thermoelectric power plants, we began to collect data on solar radiation on the island of S. Tomé. The points chosen to collect data on radiation on the island, with the help of the website (photovoltaic geographical information system),[6] were the following towns: Neves, Guadalupe, São Tomé, Santana and São João dos Angolares.

The town of S. dos Angolares has the highest radiation on the island, with an average of 200.11 Wh/m^2 or 1753 kWh/m^2. In the remaining towns, radiation is much lower, not reaching 1000 kWh/m^2, as can be seen in Table 5.[7]

3.4 Applied System Approach

We carried out 5 simulations for the city of São dos Angolares and 5 for the average of the five cities selected, with a power per panel of 200 W/m^2, we considered 400 W panels and each panel is 2 m^2. In other words, the power per area is 200 Wp/m^2.

[5] Full table in Excel—Annex I, sheet: Annual Consumption.

[6] Source: https://re.jrc.ec.europa.eu/pvg_tools/en/#MR.

[7] Full table in Excel—Annex I, sheet: Radiation.

Table 5 Radiation records on the island of S. Tomé

	Neves			Guadalupe		
Latitude (decimal degrees):	**0.355**			**0.382**		
Longitude (decimal degrees):	**6,553**			**6,639**		
Elevation (m):	**27**			**140**		
Slope	Slope: 0 deg. (optimum)			Slope: 0 deg. (optimum)		
Azimuth	Azimuth: 17 deg. (optimum)			Azimuth: 13 deg. (optimum)		
Time	Global Irr W/ m^2	Air Temp °C	10m wind speed m/s	Global Irr W/ m^2	Air Temp °C	10 m wind speed m/s
20150101:0006	0	23.75	2.36	0	22.68	2.37
20150101:0106	0	23.22	2.24	0	22.14	2.46
20151231:2206	0	24.24	1.16	0	23.05	2.12
20151231:2306	0	24.24	1.16	0	23.05	2.12
Average (W/m^2)	**111.84**	**23.38**	**1.97**	**111.73**	**22.60**	**3.38**
Total annual (kWh/m^2)	**979,70,205**			**978,75,155**		
	São Tomé			Santana		
Latitude (decimal degrees):	**0.335**			**0.26**		
Longitude (decimal degrees):	**6,721**			**6.74**		
Elevation (m):	**31**			**45**		
Slope	Slope: 0 deg. (optimum)			Slope: 0 deg. (optimum)		
Azimuth	Azimuth: 69 deg. (optimum)			Azimuth: 246 deg. (optimum)		
Time	Global Irr W/ m^2	Air Temp °C	10 m wind speed m/s	Global Irr W/ m^2	Air Temp °C	10m wind speed m/s
20150101:0006	0	25.65	3.38	0	26	3.38
20150101:0106	0	25.12	3.48	0	25.47	3.48
20151231:2206	0	26.12	3.42	0	26.51	3.42
20151231:2306	0	26.12	3.42	0	26.51	3.42
Average (W/m^2)	**103.08**	**24.81**	**3.74**	**102.42**	**25.10**	**3.74**

(continued)

Table 5 (continued)

	São Tomé			Santana		
Total annual (kWh/m^2)	**902,98,231**			**897,22,645**		

	Angolares			Average of 5 cities		
Latitude (decimal degrees):	**0.137**					
Longitude (decimal degrees):	**6.65**					
Elevation (m):	**70**					
Slope	Slope: 3 deg. (optimum)					
Azimuth	Azimuth: 13 deg. (optimum)					
Time	Global Irr W/m^2	Air Temp °C	10 m wind speed m/s	Global Irr W/m^2	Air Temp °C	10 m wind speed m/s
20150101:0006	0	26.28	5.03	0	24.87	0
20150101:0106	0	25.76	5.14	0	24.34	0
20151231:2206	0	26.82	3.94	0	25.35	0
20151231:2306	0	26.82	3.94	0	25.35	0
Average (W/m^2)	**200.11**	**25.35**	**4.16**	**125.84**	**24.25**	**125.84**
Total annual (kWh/m^2)	**1752,92,205**			**1102.32**		**1102.32**

The efficiency of the panels is around 20%, the inverter is between 90 and 95%, the losses between the panels, if we multiply 20% (0.20) * 90% (0.90) we get the total electrical efficiency of the system of 17% (0.17). A solution with 100,000 panels, 15,000,000 panels, 200,000 panels, 25,000,000 panels and 300,000 panels was simulated for each option.

In order to dimension photovoltaic energy as a way of making the most of the area used, it was decided to implement: the agro-voltaic system, giving the possibility of using these areas for agricultural cultivation of plants suitable for growing in the shade, such as coffee and cocoa; floating panels in the reservoir for hydroelectric use, avoiding the occupation of other areas on land, use of the reservoir for recreational and leisure purposes.

The aim is to use the country's hydroelectric potential to produce hydroelectric power, taking advantage of the existing hydroelectric power stations on the islands. Sizing for 2 storage basins, with a height of 50 m between basins, and storing this hydroelectric energy to use it during the night, discharging the basin to produce electricity again with the aid of reversible turbines (Francis turbines).

An efficiency of 70% was considered for the pumped hydro plant, meaning that the total efficiency of the reversible hydro plant was considered to be 0.70.

The reservoir hydroelectric system has several factors that can influence the system's performance, which means that it doesn't have the necessary speed to adapt instantly. It would therefore be advantageous to use lithium batteries for faster functionality between energy production and consumption.

In Table 7 we can see the results obtained and conclude that the simulations for the middle area, despite having smaller reservoir areas, are not the most feasible because the values are high for the available biomass compared to the S. João dos Angolares area. The simulations for S. João dos Angolares are the most balanced between available biomass and the area of the reservoir to be implemented (Table 6).

Having said that, looking only at the simulations for the S. João dos Angolares area, it can be concluded that solution 3, with 200,000 panels, is the most viable simulation, with the justification that: in simulation 3 only 17,106,174.9 kg of the available biomass will be needed and with 14.88 hectares (ha) of area to be used; it is therefore the most cost-effective solution given the need and the availability of the available resources.

Table 6 Summary of simulations

Simulation	N° of panels	S. João dos Angolares			Average		
		Costs					
		Photovoltaic system installation price (€)	Primary need for biomas (€)	Reservoir area (ha)	Photovoltaic system installation price (€)	Primary need for biomas (€)	Reservoir area (ha)
1	100 000	12,800,000	6,753,707	5.50	12,800,000	8,558,009	2.94
2	150 000	19,200,000	4,550,616	8.88	19,200,000	6,898,620	4.37
3	200 000	25,600,000	2,440,253	14.88	25,600,000	5,438,180	5.72
4	250 000	32,000,000	616,518	36.73	32,000,000	4,073,559	7.58
5	300 000	38,400,000	97,622	36.73	38,400,000	2,752,374	9.35

Table 7 Simulation result 3

Max. pumping	50 326	Turbine power	Reservoir height	50.00	m
Max. biomass	25 269	Burner power	Massof water needed	2 975 543 325,4	kg
Max. hydro	25 201	Pump power	Volume	2 975 543,33	m^3
Primary need for biomass	76 977 787	kWh	Water height (average water height in the reservoir)	20.00	m
Primary need for biomass	2 440 253	Annual biomass price	Basin area	14.88	ha
Primary need for biomass	17 106 175	kg	Negative values	–	

Table 8 Simulation result 4

Max. pumping	67 933	Turbine power	Reservoir height	50.00	m
Max. biomass	24 204	Burner power	Mass of water needed	7 346 938 776	kg
Max. hydro	25 218	Pump power	Volume	7 346 938,78	m3
Primary need for biomass	19 448 063	kWh	Water height (average water height in the reservoir)	20.00	m
Primary need for biomass	616 518	Annual biomass price	Basin area	36.73	ha
Primary need for biomass	4 321 792	kg	Negative values	–	

If we compare simulation 3 with the results of simulation 4, we see that with 250,000 panels the biomass requirement is 4321792 kg and, as a result, the hectares increase to 36.73 for the basin area to be installed (Table 8).

4 Implementation on S. Tomé Island

At this stage of the project, the advantages of implementing the project for the country will be presented, given its current energy situation. Based on the current economic situation, a pre- and post-project economic analysis will be presented.

There will also be an approach to the environmental situation before and after the project, as well as a presentation of the project's economic and environmental gains.

> A green economy is one that results in improved human well-being and social equality, while significantly reducing environmental risks and ecological scarcity. It is low-carbon, resource-efficient and socially inclusive.[8]

4.1 The Country's Economic Situation

The São Toméan economy is currently facing challenges that are typical of small, island states and that affect its capacity for autonomy. The country's economy is essentially based on agriculture, and cocoa and coffee continue to be the mainstays of the country's economy. In recent years the country has found tourism to be an essential factor for the economy, given the island's natural wealth and its extremely favorable geographical position for tourism.

[8] Source: United Nations Environment Program, 2011.

Table 9 Conversion factors and price per liter of fuel

Bus	Density kg/l	PCS (kWh/kg)	Price €
Diesel	0.846	12.67	1
Gasoline	0.737	12.89	1.2
Biomass	0.701	4.5	0.143

Table 10 Price per watt (€)

Photovoltaic panels	Inverter	Protection, wiring, support and installation	Total
0.2	0.05	0.07	0.32

Gross Domestic Product (GDP) grew at an average rate of 4.5% between 2010 and 2018, but this growth has been slowing since 2014. Economic growth was further affected in 2018 and 2019 by fuel and energy shortages, delays in government payments to local suppliers and the withdrawal of domestic financing. The negative shocks that began in 2018 continued to affect the economy's performance in 2019. The real GDP growth rate is estimated to have fallen to 2.4% in 2019, down from 2.7% in 2018. Agriculture and fisheries have been affected by climate shocks, agricultural pests and fuel and energy shortages. STP is expected to suffer a serious economic crisis as a result of the drop in tourism due to the Covid-19 pandemic.[9]

4.2 Managing the Country's Current Investment

For calculation purposes, the density of diesel was 0.846 kg/l, the calorific value was 12.67 kWh/kg and the price in euros was €1 per liter, which is equivalent to 24 doubles in STP. With regard to fuel prices, it should be noted that the price shown in Table 9 is the retail price and not the price at which the government of São Tomé buys fuel from Angola.

For gasoline and biomass, the conversion factors and prices can also be found in Table 9.

Table 10 shows the price considered when sizing the panels. We considered 0.2 cents for each Watt, 0.5 cents for the inverter that converts the electrical energy generated by the panels, and 0.7 cents for the installation of the protection system, cabling, supports and installation services, giving a total of 32 cents.

Table 11 shows the amounts currently invested in the country to produce diesel-based energy, without taking into account lubricants and wear and maintenance parts. The country currently spends 51,407,219.16 million euros per year on imports of fuel for transportation (Table 12) and the plant (the total cost of the thermoelectric plant plus the total cost of transportation).

[9] Source: World Bank in São Tomé and Príncipe, 2020.

Table 11 Cost of energy production

Thermoelectric	
Diesel consumption (litres)	26,193,958,0
Total cost (€)	26,193,958
Cost per kWh used (€) price per watt	0.269

Table 12 Fuel costs for transportation

Transports	
Diesel consumption (litres)	13,110,438
Total cost (€)	13,110,438
Gasoline consumption (litres)	10,085,686
Total cost (€)	12,102,823
Transport total cost (€)	**25,213,261**
Custo total com combustíveis (€/per year) Not getting into line with lubricant bills	**51,407,219**

4.3 Environmental Impact

In STP, *Law no. 10/99 - (D.R.) no. 15 of 31/12/1999* - Basic Law on the Environment and *Decree no. 37/99, (D.R.) no. 12 of 3/08/99* for environmental impact assessment are applied. These two pieces of legislation introduced the concept of polluter pays, consumer pays and environmental protection.

The energy and transport sectors are the biggest sources of pollution in the country, releasing pollutants such as: sulphur oxides (SO_x), ozone oxides (NO_x), carbon monoxide (CO), carbon dioxide (CO_2), volatile organic compounds (VOC), particulate matter (PM).

Looking at the situation regarding CO_2 and methane emissions on the island, according to the National Adaptation Action Plan (PANA), STP emits[10] 568,663.87 tCO2eq.

Based on the liters of fuel used as a reference for calculation purposes, it is estimated that thermoelectric power stations, together with the transport sector, currently emit 842,237,065 g of CO_2 per year. Table 13 shows the emissions of the sectors separately.

These emissions have a direct impact on the daily lives of the country's population, as the thermoelectric plant is located in the city center on the island of São Tomé. With regard to transportation in STP, there is still no age limit for vehicles in circulation and vehicles are aged, with an average age of between 20 and 25 years, thus leading to an increase in emissions.

[10] Source: National ALER status report, 2020.

Table 13 CO_2 emissions for the thermal power plant and transport

Transport diesel consumption	13,110,438	Thermoelectric diesel consumption	26,193,958
Consumption per 100 km (litres)	7	Consumption per 100 km (litres)	7
Distance (km)	150	Distance (km)	150
Travelled distance/km	280,937,965	Travelled distance/km)	561,299,100
Annual emission (gCO2)	421,406,948	Annual emission (gCO_2)	841,948,650

5 Reflection and Conclusions

This final chapter will present the ideas that remain to be realized, a conclusion in the form of a final reflection on the project, as well as the sources used to carry it out.

"It is estimated that 18–30 billion dollars a year will be needed over the next two decades for climate action and adaptation in Africa, with almost 1 trillion in investments and projects ready to be financed."—Economic Report on Africa 2020.[11]

5.1 *Final Reflection*

In Portugal, each inhabitant is expected to have their own photovoltaic panel, with decentralized and local production. In São Tomé, this option would be more difficult to implement because production is state-owned and the structures are not designed for this type of operation.

As STP is a developing country, the formatting of the energy market will be a good bet to make the market more attractive to international investors, to develop new business areas, strengthening the competitiveness of companies, as well as mitigating the costs associated with thermal power plants and offering energy security to the population.

The Portuguese Water Partnership Project report highlighted potential sites, in addition to existing ones, such as the Contador River and Guégue power stations. The same report states that STP benefits from ideal conditions for hydroelectric production. The island has relief, heavy rainfall and usable waterfalls within a radius of less than 20 km from the main towns, making the island's hydroelectric potential known. This study inventories 34 sites for a production of 244,000,000 kWh.[12] Of the 34 potential sites identified, Table 14 identifies these potential areas.

[11] Source: https://www.uneca.org/.

[12] Source: P3LP Project: Portuguese Water Partnership, water cluster, based on: *UNDP-GEF, PRODOC in CECI engineering consultants, Taiwan.*

Table 14 Places with potential for hydroelectric power production

No	Location	River	Installed capacity (MW)	Head (m)	Estimated annual production (MWh)
1	Cruz Grande	D'Ouro	0.88	100	3,461
2	Agostinho Neto	D'Ouro	0.34	60	1,340
3	Almeirim	Água Grande	0.44	50	1,731
4	Santa Luzia	Manuel Jorge	1.15	380	4,746
5	Santa Clara	Manuel Jorge	0.89	190	3,667
6	Mato Cana	Abade	2.0	60	5,599
7	Claudino Faro	Abade	2.0	100	5,348
8	Bombaim	Abade	4.0	280	9,685
9	Dona Eugénia	Io Grande	9.6	80	30,448
10	Meteus Sampaio	Umbugu	0.5	28	1,519
11	Neves	Provoz	2.0	95	7,287
12	S. João	Contador	0.9	200	1,382
13	Santa Irene	Lemba	3.0	100	9,229
14	Monte Verde	Xufexufe	0.80	60	2,935
15	Monte Rosa	Quija	3.75	260	10,427
16	Caldeiras	Carvao	0.02	50	100

6 Conclusion

The implementation proposal analysed and verified the real potential for growth and access to electricity in STP, which will be an important factor in determining the speed of the country's development.

It was concluded that with access to new technologies for energy generation, electricity can be generated at lower costs, with great use of the areas used and with great return on investment.

It is important to point out that, based on the current pattern of electricity production on the island, it is not the right way to achieve access to clean and sustainable energy. Finally, it is valid to state that after removing the current existing limitations, we believe that with the improvements and solutions proposed in this study it will be possible to help chart a path, provide more sustainable energy and meet the growing energy needs on the island, with renewable energies representing the fastest and most economical way to meet current energy needs.

Although losses currently stand at 28.6%, we believe that this is a problem that will be improved. In the solutions presented, we only considered 10% of network losses. The project will lead to an improvement in the transport sector, which translates into a positive impact for the whole island.

Based on Eco-innovation,[13] the proposals submitted will contribute to innovation on the island that aims to make significant demonstrable progress in achieving the goals of sustainable development, reducing environmental impacts, increasing resilience to environmental pressures and using natural resources more efficiently and responsibly.

It is important to note that by implementing the above-mentioned proposals it is possible to lower the island's emission rates, since the biomass plant will not contribute to an increase in atmospheric emissions, and the impact of the plant's operation on the area's air quality will be negative and negligible, since biomass is CO neutral2 and photovoltaic energy is emission-free, when accompanied by a good environmental impact assessment for its implementation.

References

1. Borhanazad H, Mekhilef S, Saidur R, Boroumandjazi G (2013) Potential application of renewable energy for rural electrification in Malaysia. Renew Energy 59:210–219
2. McCabe A, Pojani D, van Groenou AB (2018) The application of renewable energy to social housing: a systematic review. Energy Policy 114:549–557
3. Tan KM, Babu TS, Ramachandaramurthy VK, Kasinathan P, Solanki SG, Raveendran SK (2021) Empowering smart grid: a comprehensive review of energy storage technology and application with renewable energy integration. J Energy Storage 39:102591
4. UNEP P d (2015) Energy profile of São Tomé and Príncipe. Accessed 01 May 2020

[13] Source: Green Economy and Innovation Brochure.

Torrefaction and Oxidative Carbonization of Pine Sawdust and *Chlorella Vulgaris*

Bruna Petry and Margarida Gonçalves

Abstract Torrefaction is one of the thermochemical biomass conversion processes that has become popular due to its promising ability to improve the properties of biomass fuels, which can be carried out in oxidative an non-oxidative conditions. In this work, the torrefaction of pine sawdust and *chlorella vulgaris* biomass was studied at temperatures form 200 °C to 400 °C for 1 h, in batch mode, under oxidative and non-oxidative conditions. The yield of biochar produced, ash content, elemental composition and calorific value were determined in order to indicate the bahvior of these two different types of biomass under these two conditions. As a result, it was possible to conclude that temperature and the presence of oxygen have a determining influence on the properties of biomass, and that the presence of oxygen promotes an intensification of reactions during torrefaction.

Keywords Oxidative torrefaction · Oxidative carbonization · *Chlorella vulgaris* · Pine sawdust

1 Introduction

Thermochemical biomass conversion processes have gained increasing attention due to their potential to upgrade biomass and other wastes to solid, liquid and gas fuels. Among these processes, torrefaction has become an increasingly popular pretreatment due to its promising ability to improve the fuel properties of biomass, namely its apparent density and heating value. Typically, torrefaction and carbonization take place in oxygen poor atmospheres but recently torrefaction of biomass in an oxygen-rich atmosphere has been proposed as an alternative and designated as oxidative torrefaction [1, 2]. This process has been studied at temperatures from 200 °C to

B. Petry (✉) · M. Gonçalves
VALORIZA—Centro de Investigação para a Valorização de Recursos Endógenos, Instituto Politécnico de Portalegre, Portalegre, Portugal
e-mail: b.petry@campus.fct.unl.pt

MEtRICs—Mechanical Engineering and Resource Sustainability Center, NOVA School of Science and Technology, Caparica, Portugal

P. S. Brito et al. (eds.), *Insights into Energy Trends*, Green Energy and Technology,
https://doi.org/10.1007/978-3-031-83811-8_13

300 °C, with variable oxygen concentrations. The results show that oxidative torrefaction generally has higher efficiency for deoxygenation, dehydrogenation than non-oxidative torrefaction, reducing mass and energy yields but improving the biochar energy density [1, 2].

Felix and co-workers studied oxidative torrefaction of microalgae *Chlorella sorokiniana* and recommended that the conversion should be conducted at 200 °C, 10.6 min, 12% O_2 in order to maximize energy yield and the process enhancement factor [3]. The oxidative torrefaction of biomass wastes has been proposed as an efficient method to produce biochar with characteristics adequate for soil amendment [4].

One of the advantages of oxidative torrefaction is to avoid the use of a nitrogen atmosphere, but both temperature and oxygen concentration should be kept at moderate values to avoid excessive oxidation and significant mass and energy loss [5].

In this work, the torrefaction of pine sawdust (a waste from wood processing industry) and of *Chlorella vulgaris* biomass was studied at temperatures from 200 °C to 400 °C, for 1 h, in batch mode, under non-oxidative and oxidative conditions. During oxidative torrefaction the sample was exposed to atmospheric air (21% oxygen) but only in contact with the biomass surface, without imposing a continuous flowrate. The biochar yield, ash content, elemental composition and heating value were determined in order to determine the behavior of these two different biomass types under oxidative and non-oxidative conditions.

2 Material and Methods

To carry out this study, torrefaction and carbonization tests were carried out on pine sawdust and *Chlorella vulgaris* biomasses under oxidative and non-oxidative conditions. The samples (5 g) were heated at temperatures of 200 °C, 250 °C, 300 °C and 400 °C, for 1 h, in open or closed crucibles, using a muffle oven (Ceraminox, Portugal). The obtained biochars were characterized for mass yield, ash content, elemental analysis and low heating value (LHV).

The biochar mass yield represents the ratio between the masses of biochar mass and of raw biomass and is expressed in (%). The ash content was determined based on standard EM 14775:2009, with complete combustion of 1.5 g of sample in porcelain crucibles at a temperature of 550 ± 10 °C for 150 min. The elemental analysis was determined using a CHNS Analysis by FlashEA 1112 Series Analyzer (ThermoFinnigan – CE Instruments). The determination of LHV was based on the empirical formula proposed by [6] and presented below, where m_C, m_H and m_O were the contents of carbon, hydrogen and oxygen, respectively, on dry basis. When the equation applied to solid fuels, ΔH_1 should be 0.62 kJ/g.

$$LHV\left[kJ/g\right] = 38.2m_C + 84.9(m_H - m_O/8) - (\Delta H_1)$$

3 Results

The biochars produced by torrefaction under oxidative and non-oxidative conditions presented homogeneous appearance and dark colour indicating that the conversion process was effective in all conditions tested. Temperature and oxygen concentrations had significant effects in the biochar yields and properties.

3.1 Mass Yield and Ash Content

The biochar mass yields obtained for each of the temperatures in oxidative and non-oxidative conditions for pine sawdust and *Chlorella vulgaris* biomass are presented in the Table 1.

As seen in the table above, reaction temperature has a notable effect on mass yield. With increasing temperature, the biochar yield decreases for both biomasses but more but more markedly in oxidative conditions, and in both conditions for sawdust.

At 300 °C and 400 °C the biochar produced from pine sawdust in air atmosphere is less than 15% of the original sample indicating that the sample is extensively oxidized with significant loss of its carbon content. On the other hand, at 200 °C, the biochar mass yield was high and practically the same for both conditions, probably because even hemicellulose and cellulose were not decomposed at this temperature. However, when the temperature increased, the mass loss was accelerated as the reaction was intensified and more volatiles were released. The increase in temperature intensifies the thermal degradation process of hemicellulose and cellulose, and the decomposition of organic functional groups, which justifies this significant decrease in the mass yield of the solid product, which reaches 24.91 ± 0.51% in non-oxidative condition, being enhanced in the presence of oxygen, which reaches a yield of 0.92 ± 0.48%.

The mass yield of *Chlorella vulgaris* biochar shows a more gradual reduction with increasing temperature, both for the oxidative and non-oxidative conditions. At lower temperatures, it is observed that lignocellulosic biomass presents better yield,

Table 1 Mass yield (%) of biochars obtained from pine sawdust and *Chlorella vulgaris* biomass, at temperatures of 200 °C, 250 °C, 300 °C e 400 °C in oxidative and non-oxidative conditions

T (°C)	Condition	Pine Sawdust	*Chlorella vulgaris*
200	Oxid	87.21 ± 1.00	61.19 ± 0.69
	Non oxid	87.87 ± 1.52	83.73 ± 2.82
250	Oxid	16.11 ± 2.79	57.66 ± 1.01
	Non oxid	75.54 ± 5.70	69.86 ± 3.87
300	Oxid	12.15 ± 1.17	44.55 ± 1.01
	Non oxid	42.20 ± 5.65	55.00 ± 5.55
400	Oxid	0.92 ± 0.48	28.12 ± 2.37
	Non oxid	24.91 ± 0.51	34.24 ± 1.72

however, with increasing temperature, microalgae biomass presents better behavior. A review on torrefaction pretreatment indicted that differences in thermal degradation of microalgae become more significant compared to lignocellulosic types in the mild pyrolysis range. This resulted from their inherent differences in their respective components—cellulose, hemicellulose and lignin for lignocellulosic compared to carbohydrates, proteins and lipids in microalgae, both with different chemical bonding and thermal decomposition mechanisms [1, 7].

Regarding the ash content of the biochars obtained from the two biomasses at the different temperatures and conditions studied, the results are presented in the graph below.

From the Fig. 1 it is possible to verify that the ash content is much higher for *Chlorella vulgaris* biochars than for sawdust biochars which is to be expected since the corresponding raw materials also present a significant difference on this parameter, respectively 0.15% and 9.5%. For both biomasses the oxidative condition was always more severe than the non-oxidative one, which was expected considering that the presence of oxygen, resulting in a higher oxidation extent and a consequent enrichment of the ash fraction. This characteristic of the biochars obtained in oxidative conditions can be an advantage for soil amendment applications [4].

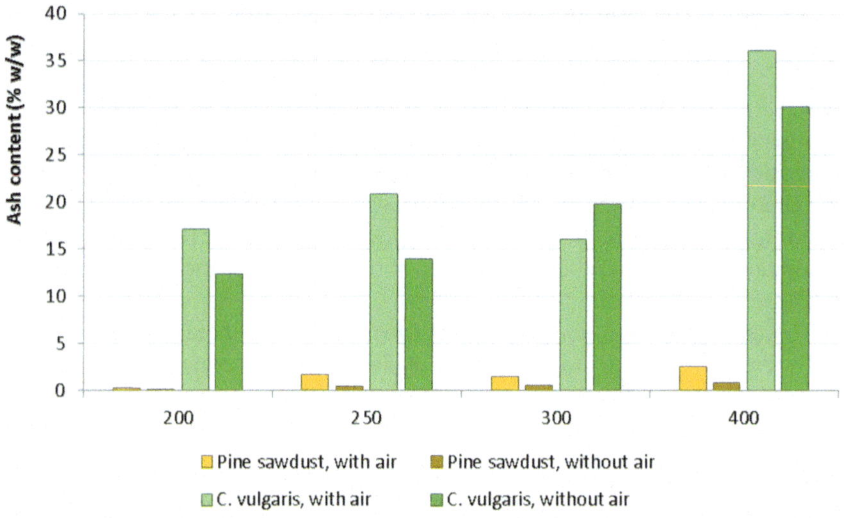

Fig. 1 Ash content of the biochars obtained from pine sawdust and *C. vulgaris* biomass, at the different conditions studied

3.2 Ultimate Analyses and Heating Value

The results of the elemental analysis (carbon, nitrogen, hydrogen and phosphorus content) of the biomasses and conditions studied are presented in Table 2.

Using the elemental composition of the produced biochars, the respective atomic H/C and O/C ratios were determined and represented in the Fig. 2 using a Van Krevelen diagram. These atomic rations reflect the aromatization and deoxygenation reactions that took place during the torrefaction processes.

The H/C ratios and O/C ratios, important indicators for evaluating the performance of energy density in biomass torrefaction [8] of the biochars decrease with temperature for both conditions, mainly as a result of the decrease of hydrogen

Table 2 Elemental composition of pine sawdust and *Chlorella vulgaris* biomasses from the different conditions studied

Condition	T (°C)	Pine sawdust				*Chlorella vulgaris*			
		N	C	H	S	N	C	H	S
Raw biomass		0.15	45.90	6.35	0.00	8.76	45.83	6.85	0.52
Oxidative	200	0.16	48.19	6.17	0.00	10.06	54.13	4.02	0.38
	250	0.37	63.84	2.92	0.00	9.83	52.28	4.09	0.14
	300	0.37	67.01	2.73	0.00	8.78	56.89	4.08	0.20
	400	0.46	51.52	1.91	0.00	6.86	48.13	1.96	0.25
Non-oxidative	200	0.15	49.96	6.01	0.00	9.79	51.98	5.64	0.45
	250	0.17	53.45	5.70	0.00	9.70	54.54	5.48	0.00
	300	0.26	67.74	4.42	0.00	9.27	58.18	5.26	0.20
	400	0.30	76.73	3.17	0.00	7.85	53.20	3.05	0.17

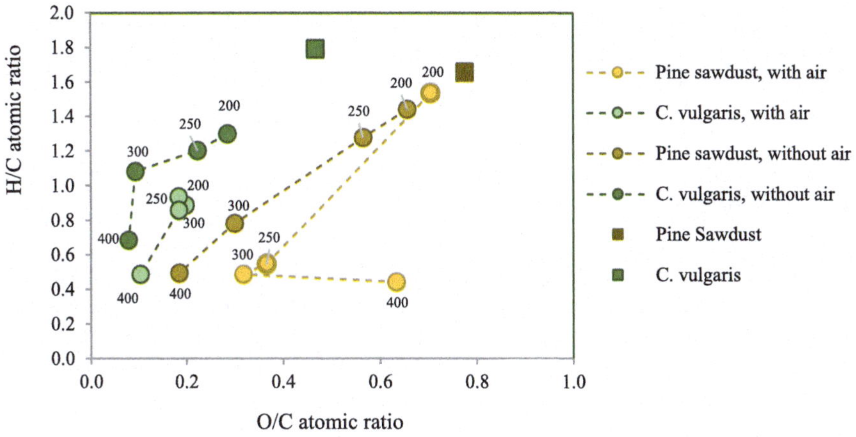

Fig. 2 Van Krevelen diagram of the O/C and H/C ratios of the different biochars produced in the conditions studied

and oxygen concentrations while carbon concentration tends to increase in the biochars. Hydrogen content of the biochars drops more sharply from 200 °C onwards and continues to decrease with increasing temperature for both conditions thus corresponding to more aromatic structures.

Considering the carbon content of raw pine sawdust biomass (45.9%), it was possible to observe that the carbon content increases with temperature reaching 67.0%, at 300 °C in an oxidative environment, and reaching 76.7% at 400 °C in non-oxidative environment. On contrary, oxygen decreases from 47.5% to 28.4% for biochar produced at 300 °C in air atmosphere and 18.9% for biochar produced at 400 °C in non-oxidative conditions. The effect of excessive carbon oxidation can be observed at 400 °C, not only by the increase in ash content but also in the increase of O/C ratio.

The effect of the feedstock composition in the final characteristics of the biochars can be observed in the different behavior of *Chlorella vulgaris* biomass in comparable conditions. The carbon content of the *Chlorella vulgaris* biomass increases to 56.9% for biochar produced at 300 °C in oxidative conditions and to 58.2% for biochar produced at the same temperature in non-oxidative conditions. Oxygen concentration of *Chlorella vulgaris* biomass (28.5%) is already lower than the oxygen concentration of pine sawdust and further decreases with temperature in both conditions. The biochar produced at 300 °C in oxidative conditions had 6.7% oxygen and the biochar produced at the same temperature in non-oxidative conditions had 5.6% oxygen.

This relationship of increasing carbon content and reducing hydrogen with increasing temperature may occur due to the decarboxylation and decarbonization of hemicellulose and cellulose accompanied by the release of water, oxygenated carbohydrates, in addition to the carbonization of some volatile materials that are released during the torrefaction.

Using data obtained from elemental analysis and ash content, LHV values were estimated, which are presented in the Fig. 3.

For the temperatures of 250 °C and 300 °C the biochars produced in both conditions reached LHVs above 20 MJ/kg. Nevertheless, at 400 °C excessive oxidation results in a degradation of the heating values and mass yields, especially for lignocellulosic biomass in oxidative conditions.

4 Conclusions

Through this study it was possible to conclude that the torrefaction temperature and the presence of oxygen has a notable influence on the properties of the biomass product. The presence of oxygen ends up intensifying and anticipating reactions that occur in non-oxidative conditions. Among the conditions studied, oxidative torrefaction can be used at relatively low temperatures and when the formation of biochars with high aromaticity and ash content is the intended effect.

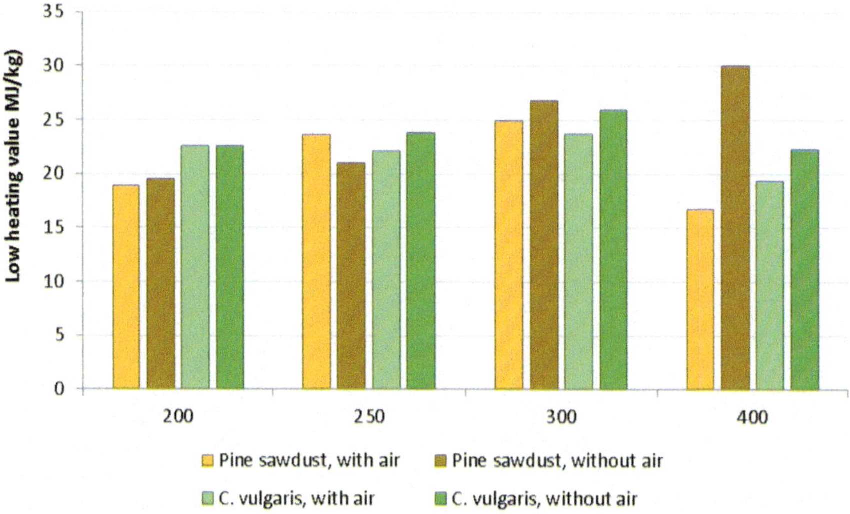

Fig. 3 LHV of the biochars produced from pine sawdust and *C. vulgaris* biomass by oxidative and non-oxidative torrefaction

References

1. Brachi P, Chirone R, Miccio M, Ruoppolo G (2019) Fluidized bed torrefaction of biomass pellets: a comparison between oxidative and inert atmosphere. Powder Technol 357:97–107. https://doi.org/10.1016/j.powtec.2019.08.058
2. Chen D, Chen F, Cen K, Cao X, Zhang J, Zhou J (2020) Upgrading rice husk via oxidative torrefaction : Characterization of solid, liquid, gaseous products and a comparison with non-oxidative torrefaction. Fuel 275(April):117936. https://doi.org/10.1016/j.fuel.2020.117936
3. Felix CB, Chen W, Chang J, Park Y, Saeidi S, Kumar G (2023) Oxidative torrefaction of microalgae Chlorella sorokiniana: process optimization by central composite design. Bioresour Technol 382(May):129200. https://doi.org/10.1016/j.biortech.2023.129200
4. Thengane SK, Kung KS, Gupta A, Ateia M, Sanchez DL, Mahajani SM, … Ghoniem AF (2020) Oxidative torrefaction for cleaner utilization of biomass for soil amendment. Clean Eng Technol 1(December):100033. https://doi.org/10.1016/j.clet.2020.100033
5. Sui H, Chen J, Cheng W, Zhu Y, Zhang W, Hu J, … Chen H (2024) Effect of oxidative torrefaction on fuel and pelletizing properties of agricultural biomass in comparison with non-oxidative torrefaction. Renew Energy 226(March):120423. https://doi.org/10.1016/j.renene.2024.120423
6. Hosokai S, Matsuoka K, Kuramoto K, Suzuki Y (2016) Modification of Dulong's formula to estimate heating value of gas, liquid and solid fuels. Fuel Process Technol 152:399–405. https://doi.org/10.1016/j.fuproc.2016.06.040
7. Zhang C, Chen W, Ho S, Zhang Y, Lim S (2023) Comparative advantages analysis of oxidative torrefaction for solid biofuel production and property upgrading. Biores Technol 386(July):129531. https://doi.org/10.1016/j.biortech.2023.129531
8. Zhang C, Wang C, Cao G, Chen W, Ho S (2019) Comparison and characterization of property variation of microalgal biomass with non-oxidative and oxidative torrefaction. Fuel 246(December 2018):375–385. https://doi.org/10.1016/j.fuel.2019.02.139

The manufacturer's authorised representative in the EU is Springer Nature Customer Service Centre GmbH, Europaplatz 3, 69115 Heidelberg, Germany. If you have any concerns regarding our products, please contact ProductSafety@springernature.com

Printed and bound by CPI Group (UK) Ltd, Croydon, CR0 4YY
30/09/2026
02225604-0010